Math for Girls and Other Problem Solvers

Diane Downie, Twila Slesnick, Jean Kerr Stenmark

Math/Science Network, Lawrence Hall of Science, University of California

The Lawrence Hall of Science is a public science center, teacher training institution, and research unit in science education at the University of California, Berkeley. For many years, it has developed curricula and teaching strategies to improve mathematics and science education at all levels, and to increase public understanding of, and interest in, science.

The Math/Science Network is an association of 800 scientists, educators, engineers, community leaders, and parents who work cooperatively to increase the number of women interested in and qualified for scientific and technical careers. The network provides support and coordination for programs designed for students of all ages, educators and parents, women resuming scientific work, and professional scientists and engineers.

The Network's secondary programs are at the Lawrence Hall of Science, under the direction of Nancy Kreinberg. The Math/Science Resource Center, coordinated by Jan MacDonald, is housed at Mills College.

Credits:
Photography, Elizabeth Crews, Eileen Christelow
Illustration, Eileen Christelow
Design, Eileen Christelow

Additional copies of the Handbook are available from:

Lawrence Hall of Science
University of California
Berkeley, CA 94720
Attn: Careers

Printed in the United States of America.
7 8 9

Mayer

Contents

	• Logic Strategies and Patterns	• Breaking Set
DAY 1	Balloon Ride (Nim) Guess	Mystery Story
DAY 2	Hex People Sorting Attribute Activities Spatial Creatures	Mystery Story
DAY 3	Pentro Function Machines	Mystery Story
DAY 4	Horseshoe Game Wolf & Hounds	Topological Rope Puzzle Mystery Story
DAY 5	Leap Frog Map Colors	Mystery Story Toothpick Puzzles
DAY 6	Fiddle Faddle Flop Hurkle Tower of Hanoi	Mystery Story
DAY 7	Span Kalah Where's the Rectangle?	Mystery Story
DAY 8	Dougle Digit Multi-Color Sticks	Mystery Story

Strands Chart

• Creative Thinking • Estimation • Observation	• Spatial Visualization	• Careers
Name Game Class Age Blind Circles	Blind Circles	Jobs Men & Women Do
Survey & Graph	Hex Spatial Creatures	Who Am I?
Man in Pit Fictionary	Pentominos Pentro	Career Skits
Mystery Person Geoblocks Double Design	Geoblocks Topological Ropes Double Design	
How Many Frogs? Leap Frog	Find Your Way Home Coordinate Dice Toothpick Puzzles Map Colors	Which Job Do I Have?
How Many Beans?	Build A Structure Hurkle Tower of Hanoi	This Is Your Life
Guess the Carob Balls Inventions	Origami Box Where's the Rectangle?	
Double Digit	Symmetric Art Build a Puzzle Multi-Color	Collage of Women Working

Foreword

Math for Girls was born on the day in 1973 when Berkeley sociologist Lucy Sells came to the Lawrence Hall of Science to inquire what was being done to reverse the pattern of math avoidance she was observing in women college students. Because a significant number of these women were coming to U.C. Berkeley with insufficient mathematics preparation to enter the majority of majors on campus, she reasoned that the problem needed to be addressed much earlier than the high school years.

As a public science center, the Lawrence Hall offers after-school classes in the physical and life sciences, mathematics, and computer science for children of all ages. A look at the enrollment of these classes showed that less than 25 percent of the students were girls. In other words, parents were sending their sons to learn how to program a computer, experiment with motors, design an animal habitat, or explore mathematical puzzles and patterns, but they were not providing their daughters with the same opportunities.

This finding led to development of a class that would encourage girls to come to the Lawrence Hall and to enjoy learning mathematics. For many who enroll in the class, math is a subject that either scares, mystifies, or bores them. Math for Girls introduces these students to "hands-on" experiences in logical thinking and problem solving that stimulate their curiosity and interest in doing mathematics. Mathematics becomes as much fun as it is challenging.

Since the first class in Spring, 1974, over 800 girls ages six to fourteen have taken Math for Girls. The total number of women students enrolled in other Lawrence Hall classes has doubled, increasing the total female enrollment to 36 percent of the student population.

The original course curriculum was developed by Dr. Diane Resek, now Associate Professor of Mathematics, San Francisco State University. She set the goals, developed the format, and trained the teachers. Six years later the course is as active as when she created it—a tribute to Dr. Resek's skills in both mathematics and teacher education.

The course has flourished under the leadership of these creative, talented teachers: Rita Liff Levinson, Heather Hubbard, Arlene King, Barbara Jasinski, Nancy Weeks, Stephanie Johnson, and Tina de Benedictis. Diane Downie took over responsibility for Math for Girls in 1978 and has nurtured, groomed, and strengthened it with great skill and care. It has been under her tenure that a considerable number of activities, ideas, strategies, and materials have been collected, revised, refined, and presented in this Handbook.

The Math for Girls course did more than bring girls to the Lawrence Hall of Science. It was the seed for what has become a major focus at the Hall encompassing workshops for teachers, conferences for adolescent women, and programs for working women. It put Lawrence Hall staff in contact with others who were developing programs to encourage the participation of girls and women in mathematics and science.

Out of those meetings grew the Math/Science Network. This unique association of 800 educators, scientists, engineers, community leaders, and government workers has demonstrated the effectiveness of collective action focused on a critical issue. Since 1978, when the Network was officially established with a grant from the Carnegie Corporation of New York, Network members have introduced 12,000 junior and senior high school women students in Northern California to career opportunities in math and science (at 50 conferences); provided inservice training to hundreds of elementary and secondary educators and administrators; developed courses at local colleges and universities to increase students' comprehension in mathematics, statistics, and computer science; provided information and contacts to encourage 3,000 college and re-entry women to pursue science careers; and created booklets, films, and resource packets to promote the increased interest and participation of girls and women in mathematical and scientific fields of study and work.

The public has become aware that young women need to prepare for a future with the greatest number of options. It is now up to each of us, individually and collectively, to ensure that girls and young women receive the best education we can provide.

The Math for Girls Handbook is one contribution to that effort. We hope you find it useful.

Nancy Kreinberg, Director of
Elementary and Secondary Programs
Math/Science Network
Lawrence Hall of Science
University of California, Berkeley

Introduction

The content, goals, environment, and organization of this book arose naturally out of the need described in the Foreword. Although the course has been refined and polished, the basic structure and intent has remained the same over the past six years.

CONTENT

The content focus of the activities is on problem solving rather than on traditional arithmetic skills. In this class, problem solving goes beyond solving word problems through exploring unusual problems that require logical and creative thinking.

The developers of the program identified certain characteristics of good problem solvers and articulated errors commonly made during the problem solving process. They then molded their instructional approach and developed content around four problem solving skill areas and a career component.

According to Whimbey, Piaget, and a host of other educators and psychologists, a good problem solver is active physically and mentally. A list of problem-solving characteristics would include:

- drawing, visualizing, using props
- approaching the problem in a systematic or logical way
- breaking out of a thinking rut or set
- using accuracy—especially in observation
- looking at different and simpler models
- visualizing and mentally manipulating objects in space
- having confidence—willingness to risk trying new ideas

Identification of these characteristics gave rise to the following problem solving strands:

A. USING LOGIC, STRATEGIES, AND PATTERNS. Activities in this strand focus on systematic problem solving. Some fundamental and general techniques are introduced in a recreational format. Solutions to fantasy logic problems are sought, pattern-finding activities are presented, games are played, and appropriate strategies are identified.

B. BREAKING SET. These activities present a seemingly impossible problem to the student. The apparent solution path is almost always incorrect. Solution requires being open to non-obvious processes. With experience, as well as exposure to simpler models of the same problem, students can succeed with these difficult problems. Activities include topology rope puzzles, mystery stories, number patterns, and classic math problems.

C. CREATIVE THINKING, ESTIMATING, AND OBSERVING. Solving problems without a predesignated right answer leads students to use ingenuity. Problems simulate the real world in which one must seek and evaluate solutions to real problems. The activities require inventing, pretending, building, and experimenting.

Students are also encouraged to use observation and communication skills to help define and clarify problems and facilitate the solution process. Activities involve giving detailed directions or descriptions, constructing models, participating in non-verbal group constructions, and observing events.

D. SPATIAL VISUALIZING. Solutions to problems often are found through pictorial representations or through the construction of three-dimensional models. Sometimes, however, models cannot be constructed, in which case, the ability to analytically examine a visual representation of the problem is necessary. To develop this skill, students can examine optical illusions, symmetry in art, paper folding, rope configurations, and the effects of reflections and rotations.

In addition to the problem solving strands, there is a career component built into the course. Career awareness activities take the form of brainstorming, discussions, skits, and personal written descriptions or art work. The intent is to make students aware of their own biases and those of society. Instructors serve as role models for students. Students learn that many careers are open to them if they do not shy away from mathematics and science.

GOALS

Two important ideas, improvement of attitudes and development of skills, provided a framework for the development of the course. The primary goal is to ensure that students increase positive feelings and decrease fearful attitudes about mathematics.

A second aim of the course, to increase problem solving skills, may mean approaching problems more systematically, sharpening powers of logical deduction, seeing patterns to solve problems, being willing to try new ideas, or generating new methods of tackling a problem.

CLASSROOM ENVIRONMENT

To accomplish the above goals for seven- to thirteen-year-old students it was easy to identify a recreational milieu as most effective. Games and puzzles provide an atmosphere of fun, increasing the likelihood of lasting interest in mathematics. Activities have been created or adapted to give an environment that is cooperative rather than fiercely competitive. A non-threatening, supportive atmosphere encourages intellectual exploration and risk-taking, especially for girls.

ORGANIZATION

The activities in this book all belong to one or more of the five strands (see matrix on page 00). The course is organized according to a schedule including 8 separate days. Activities from several or all of the different strands appear in each 1½- to 2-hour class. This was considered preferable to having days with conceptual themes for several reasons. First, similar activities with similar goals presented one right after the other can be monotonous and boring to young students. Variety tends to increase interest and extend involvement time. Second, Bruner's notion of the spiralling curriculum could be implemented. Each week the students participate in activities from the different strands so that past learning is constantly reinforced and transferred to similar but not identical problems or activities. Spiralling allows a teacher to introduce a concept or skill gradually, building a strong foundation without over-loading or boring the learner.

Each activity is presented here in a format that identifies the strand to which it belongs, delineates the mathematical content, and provides a rationale for the activity. Instructions, materials, and variations are written simply and concisely to facilitate teacher preparation. Occasional reference is made to commercially available materials, and sources of these materials are given in the Appendix.

LANGUAGE

While the original course on which this book is based is called "Math for Girls," these activities are appropriate for both boys and girls. Consequently, throughout the book, we use the term "students" rather than "girls" to include all learners. We do, however, retain the pronouns "she" and "her" to refer to all students.

HOW TO USE THIS BOOK

For those who are interested in establishing and teaching a Math for Girls class in a school, museum, recreation center, or private educational organization, this manual can serve as a complete teacher's guide or as a source of ideas.

However, numerous other possibilities exist. Each day could serve as an independent 90-minute or two-hour workshop. This workshop might be presented to girls at a Scout meeting, in an after-school club, or in the classroom as part of the mathematics curriculum.

Activities might also be organized by strands if an educator should want to emphasize a specific problem solving skill area. For example, the logic, patterns, and strategy activities could be gathered and presented over a period of time as a problem solving unit.

CONCLUSION

This package of activities is designed to involve students and teachers in thinking about problem solving in new ways and forms. There are no standard problems here. The problems require mental and physical activity. They encourage creative thinking by asking for and allowing non-standard problems and solutions. Versatile thinking is a goal. It is believed that those who can solve many different types of activities are more likely to be able to handle the variety of problems encountered in the adult world.

Day 1

Name Game

- **Creative thinking**
- **Cooperation**
- **Memory strategies**

MATERIALS

- **Pencil for teacher**
- **Roll call sheets**
- **Name tags (optional)**

This activity introduces each group member in an informal, relaxed way. It also sets the tone of the class, providing an environment that encourages cooperation, creativity, and acceptance of one another's ideas.

PEOPLE: Whole class, in a circle

ACTION:

Students sit in a circle on the floor. Explain that each person will make up a name using her real first name and another word that rhymes, begins with the same letter, or expresses something about her personality. Students may give each other suggestions. Some examples are: Diane Cayenne, Magical Mary, Brave Barbara.

One of the instructors begins the introductions by sharing her made-up name. The next person says the instructor's name and then her own. As the game continues around the circle, each person repeats the name of the preceding person and then announces her name. At the end, ask for a volunteer to try to recite all the names in the group. Everyone should give lots of hints.

IF YOU LIKE:

At the end of the class, gather the group together and go through the names again. The second day should begin with a review of the names. A variation that is more difficult, but fun, is to have each person repeat the names of all those preceding her.

The Class Age

- **Estimation**
- **Arithmetic**

MATERIALS
- **Calculators**
- **Chalkboard**
- **Chalk**

Students should be in the habit of estimating to decide if their solution to a given problem is reasonable. When estimation becomes an inherent part of the problem solving process, students become less right-answer oriented and are less afraid to take risks — to make a guess that might be "wrong."

PEOPLE: Whole class

ACTION:

Have each student guess what the sum of everybody's age will be. Write the guesses on the chalkboard.

Give everyone a calculator.

Go around the group asking each person how old she is. Students enter each age in the calculator and then push "+."

Ask for the subtotal after each entry. If anyone has made an error, she can clear her calculator and start with the current subtotal.

Compare the actual total with the estimates.

IF YOU LIKE:

Estimating a group total can be done for other attributes. For example, students could estimate the combined height of the group: How long a line would they make if they lay down head to toe? This would, of course, involve measurement. Older students could estimate the *average* age or height of the group.

Balloon Ride*

- **Patterns**
- **Strategies**
- **Problem solving**

MATERIALS
- **Toothpicks**
- **Balloon Ride gameboard**

Strategy games encourage students to find patterns and techniques such as breaking a problem down into a simpler one or working backwards.

PEOPLE: Two

ACTION:

Set up the gameboard with 10 toothpicks connecting the bottom of the balloon with the ground. The object is to get a free ride by removing the toothpicks according to a strategy.

Read the story to students:

Balloon Ride

The Hot Air Balloon is coming to town. Free rides will be given to anyone who can cut the last tie rope holding the balloon to the ground. Here are the rules:

- 10 tie ropes hold the balloon to the ground.
- Two people take turns cutting ropes. Each person can cut 1 or 2 ropes on a turn.
- Whoever cuts the last rope gets a free ride.

Can you figure out how to get the free ride every time?

IF YOU LIKE:

Balloon Ride can be extended by increasing the number of tie ropes and also by increasing the number of ropes each player may cut. For example:

- 21 tie ropes; players may cut 1, 2, or 3 ropes at a time.
- 40 tie ropes; players may cut 1, 2, 3, or 4 ropes at a time.

Would you rather be the first or second player for each of these?

*This is one of many versions of the ancient Chinese game of NIM.

BALLOON RIDE

(ditto sheet)

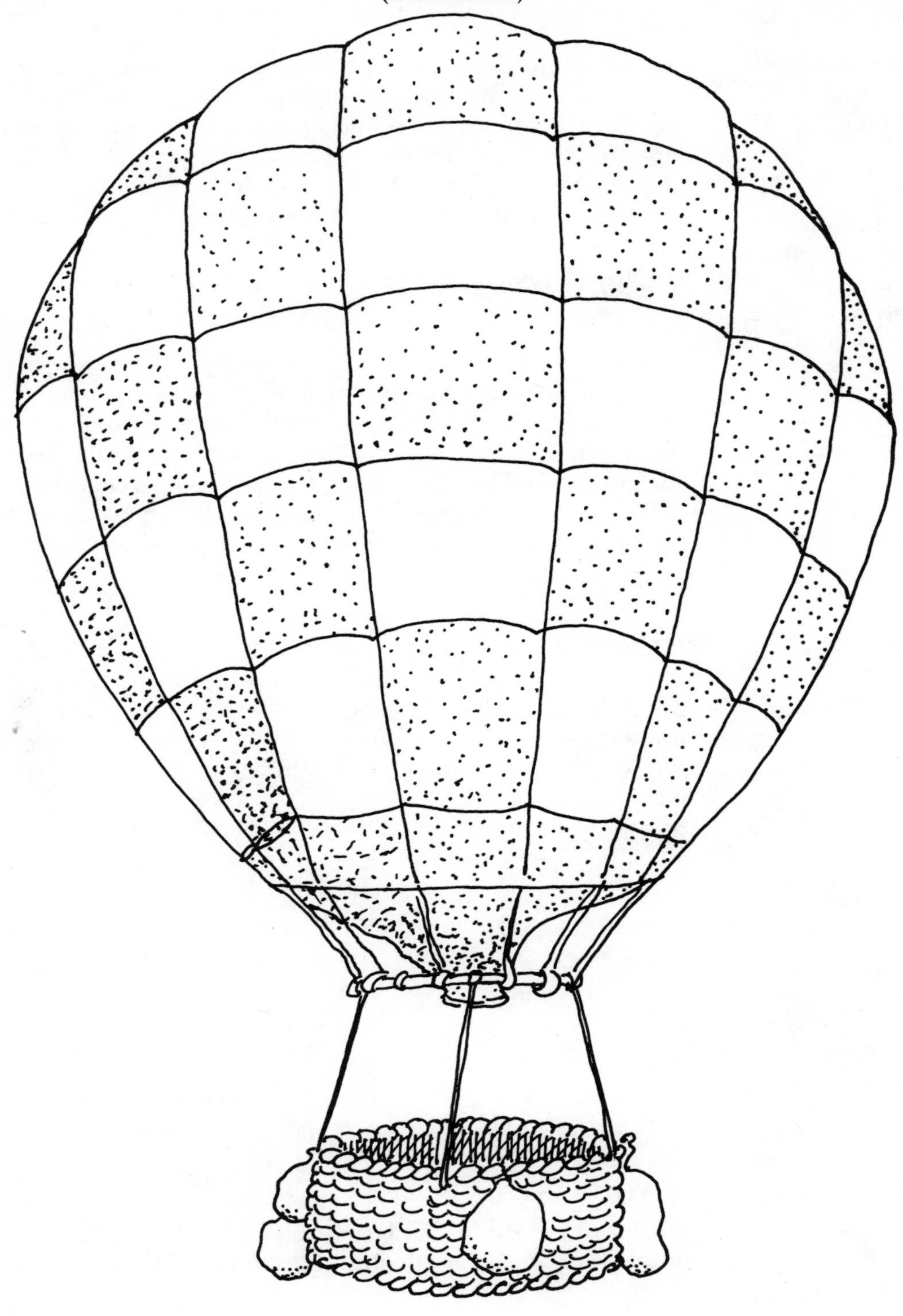

Guess

- **Binary search strategy**
- **Patterns**
- **Problem solving**

MATERIALS

- **Blackboard (or butcher paper) with the numbers 1-100 written on it**
- **Chalk (or felt pen)**

As the students guess a secret number between 1 and 100, the instructor records the number of guesses it takes to find the number. In attempting to reduce the required number of guesses, students discover strategies. Unlike many other strategy games, this one always leads to success for all participants. Consequently, the strategy can be uncovered in a positive setting.

PEOPLE: Whole class

ACTION:

Think of a number between 10 and 100. Write it on a slip of paper and put it out of sight. The students then begin to guess numbers. Tell them if their guesses are "too big" or "too small". Keep track of the guesses and clues on the chalkboard using a chart (see example below). After each guess, ask which numbers have now been eliminated. Record the number of guesses needed to find the hidden number.

Strategy:

After a few games, ask if anyone has a strategy. If necessary, you can direct the students to a strategy with questions like: How many numbers were there to start with? What guesses eliminated lots of numbers? What is the largest number of guesses we would ever need?

Example

guess	clue
17	too small
89	too big
34	too small
40	too small
50	too small
71	too big
63	too big
54	just right! (8 guesses)

IF YOU LIKE:

With younger students, play "Guess" with the numbers 1-20. Use a number line and erase the numbers that are eliminated by a guess.

Jobs Men And Women Do

This activity introduces the concept of brainstorming — the generating and accepting of any and all ideas for purposes of expanding perspective on a problem. Students also become aware of their own ideas of male and female roles and their own values. Ideally, the students will become familiar with more occupational possibilities and with a variety of careers.

- **Brainstorming**
- **Career awareness**
- **Male-female roles**
- **Values**

MATERIALS
- **Chalkboard (butcher paper)**
- **Chalk (felt pen)**
- **Index cards**

PEOPLE: Whole class

ACTION:

Indicate that the group is going to brainstorm a list of jobs they see men do and a list of jobs they see women do. Brainstorming means that everyone should mention whatever idea comes to her and it will be accepted and recorded.

Have students start generating the two lists and record all the occupations mentioned. Compare the two lists.

Ask if women can do all the jobs on the men's list and vice-versa. Why do they think so? Contrast jobs women *can* do with jobs women usually hold.

Star those jobs students are most interested in and would like to know more about. which jobs require more than two years of high school math? Talk about math courses available and how these courses will help them in their careers.

IF YOU LIKE:

Have students make occupation lists according to people they know.

The favorite occupations should be recorded and saved for future use.

Mystery Story

- **Group problem solving**
- **Breaking set**
- **Questioning strategies**

MATERIALS
- **None**

This activity stimulates creative thinking. Very often the student must break away from the solution path she has set for herself. Sometimes this can be accomplished by asking questions which lead in another direction entirely.

PEOPLE: Whole Class

ACTION:

Tell the students you are going to ask them to solve a mystery. There is one rule—they may ask only questions which have yes or no answers. Any such question, however, is acceptable.

Tell the following story: Bob, Carol, Ted and Alice all lived together in an apartment. One night Bob and Carol went out to a movie. When they got home, they found Alice dead on the floor in a pool of water with broken glass all around her. Ted was asleep on the couch. What happened?

Solution: Alice is a fish who died of suffocation when Ted, the playful pupppy, knocked the aquarium on the floor.

If the group runs out of questions, give hints about which facts are important.
For example:
- Where did the glass and water come from?
- Why were the police never called?
- Is there anything unusual about Ted or Alice?

IF YOU LIKE:

Have small groups make mystery stories. Discuss what makes the mysteries difficult to solve? What are examples of other facts that might surprise us?
- unusually tall or short person
- woman with a "man's" job
- words with double meaning, such as, *a ring* is both an object and a sound; a *bicycle* is both a vehicle and a deck of cards.

For more mystery stories, a good source is:

Weintraub, Richard and Richard Krieger
Beyond the Easy Answer 1978
Zenger Publications
Gateway Station 802
Culver City, CA 90230

Blind Circles

Many problems new to the solver seem to be nearly impossible. It often helps to think about the problem in an unusual way. This activity places students in a situation they probably have not encountered before. Group cooperation could make the difference between success and failure in solving the problem.

- **Problem solving**
- **Cooperation**

MATERIALS
- **None**

PEOPLE: Whole class

ACTION:

I. One Circle
Clear an open space in the middle of the room and have everyone gather together. Tell everyone to close their eyes. Their first task is to form one large circle keeping their eyes closed until they have completed the task and you tell them they may open their eyes.

II. Two Circles
Have the group close their eyes again, drop hands, and break the circle formation. This time ask them to form two separate circles. When they have succeeded, have them look. Without breaking up the two circles, have everyone close their eyes again. Ask one circle to get inside the other without specifying which circle goes inside the other. Then ask the two circles to trade places.

IF YOU LIKE:

Many variations of this exist. The group can be asked to make different formations, such as one long line or two parallel lines. Discuss what made the activity difficult. Why were they able to perform the tasks?

Day 2

Mystery Stories

- **Group problem solving**
- **Breaking set**
- **Questioning strategies**

MATERIALS
- **None**

This activity stimulates creative thinking. Very often the student must break away from the solution path she has set for herself. Sometimes this can be accomplished by asking questions which lead in another direction entirely.

PEOPLE: Whole class

ACTION:

Tell the students you are going to ask them to solve a mystery. There is one rule — they may ask only questions which have yes or no answers. Any such question, however, is acceptable.

Surgeon's Son

Tell the following story: A father and son are in a car accident. The father is unconscious. The son is rushed to the hospital for emergency surgery. The surgeon takes one look at the boy and exclaims: "I can't operate on this child, he's my son!" What's happening here?

Solution: No, the surgeon is not the boy's stepfather, godfather, or grandfather; she's his mother. Similar mystery stories can be invented that revolve around women in nontraditional occupations. Breaking *this* set is one of the goals of Math for Girls.

If the group runs out of questions, give hints about which facts are important.

The Case of the Ring

Tell the following story: Two people are near each other. The first reaches out, grabs the other's ring, pulls on it, and then drops it. The second person screams "Thank you!" Why was she so grateful?

Solution: They are parachutists. The second person couldn't reach the ring to open her parachute, so the first person pulled it for her.

IF YOU LIKE:

Discuss what makes the mysteries difficult to solve. What are examples of other facts that might surprise us?

- unusually tall or short person
- woman doing a "man's" job
- words with double meaning, such as, *a ring* is both an object and sound; *a bicycle* is both a vehicle and a deck of cards

Who Am I?

Students use guessing strategies to determine the area of expertise of the instructors. The students in the class find out that the instructors are women studying or working in fields that are non-traditional for women. The relationship of math to these fields may need to be discussed.

- **Career awareness**

MATERIALS
- **None**

PEOPLE: Whole class

ACTION:
Challenge the group to guess the field of study of one of the instructors in 10 guesses or less. Take questions one at a time and answer only yes or no. A second instructor can record the guesses and responses on butcher paper. This provides a record of what students believe are the most likely fields of emphasis for women. Once the academic major of the first instructor is known, the others are usually guessed more easily.

IF YOU LIKE:
Each instructor could act out some aspect of the work she will do in her chosen career.

If all of the instructors are teachers, choose some other person— a working mother, an aunt, a person in a news picture.

Hex

- **Strategy**
- **Spatial visualization**

MATERIALS
- **5 x 5 Hex boards**
- **6 x 6 Hex boards**
- **10 x 10 Hex boards**
- **Crayons**

This game gives students an opportunity to develop winning strategies very quickly. It also requires that players take many variables into account before each move. Each person must examine all the move options she has and project the possible or probable responses her opponent will make to each move.

PEOPLE: Two

ACTION:
Each pair of students begins with a 5 x 5 Hex board and two different colored crayons. One player is designated "X" and the other "O". Each chooses a crayon color. Players take turns marking an X or an O in a hexagon. The winner is the first one to complete a path connecting her two sides of the board. The corner hexagons belong to both of the two adjacent sides.

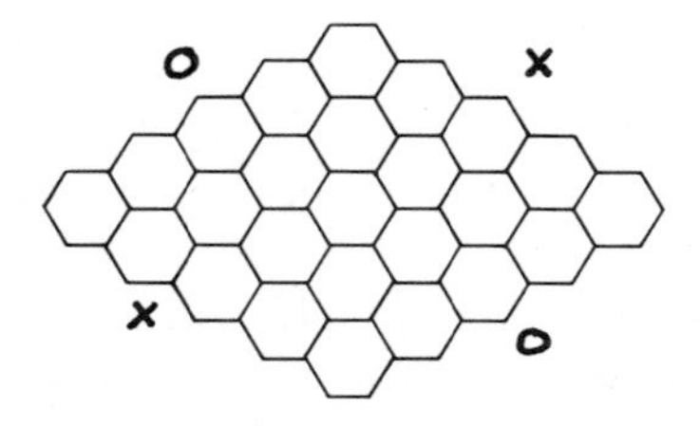

IF YOU LIKE:
Encourage students to play several games in a row on the small playing boards (25 hexagons). Let the loser of each round decide whether to play first or second on the next round.

Discuss strategies with the whole group. List on butcher paper or a blackboard the ideas generated.

Focus on questions such as:

- Did the first player win more often or did the second player win more often?
- Where's a good place to play first?
- Is there always a winner?
- Is it best to play offensively (i.e., concentrate on connecting hexagons) or defensively (i.e., try to block your opponent's moves)?

After the discussion, suggest that students play three more games (perhaps on larger boards) and try out the ideas mentioned.

(ditto sheet)

X
O
O
X

X
O
O
X

X
O
O
X

People Sorting

- **Logic**
- **Sorting**
- **Attribute identification**

MATERIALS
- **One large loop of yarn**

This activity introduces the concept of an attribute and sets the stage for development of logic skills through work with attribute blocks and spatial creatures. Everyone gets involved.

PEOPLE: Whole class

ACTION:

Put a large loop of yarn on one side of the room. Name a characteristic such as "wearing something green" and have all those who fit that description stand inside the loop. Repeat the activity with several other characteristics such as: having green eyes, wearing a belt, liking dogs. Let the students think of some of the characteristics.

Think of a secret characteristic. Without telling the characteristic, the leader has each person with that characteristic step inside the loop. The challenge is for the group to guess the mystery attribute.

Ask the group what attributes are shared by everyone in the room.

IF YOU LIKE:

Seated in a circle, ask each person to name an attribute she has that is different from anyone else in the group. Point out when appropriate that to say, "I have glasses," might not be enough to distinguish one person from everyone else. Several people may be wearing glasses.

Have one student leave the room. The group sorts itself according to any rule chosen by the group. The "Outsider" comes in and tries to guess the rule.

Attribute Activities

- **Attributes**
- **Sorting**
- **Classification**

Attribute activities encourage the observation of differences. They provide experience with classification and logical reasoning tasks which help develop systematic thinking. The activities range in difficulty from very simple to challenging.

- **MATERIALS**
- **Attribute shapes. A kit consists of yarn loops, labels, and 32 shapes; 4 different shapes, 2 different sizes, and 4 different colors. (See page 32 for patterns). Attribute blocks may also be purchased commercially.**
- **Attribute labels. A set consists of 10 1" x 2" cards:**

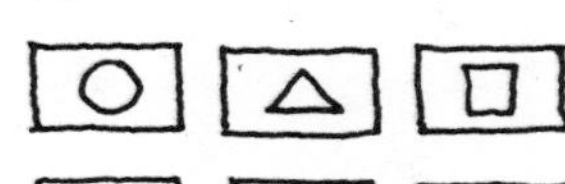

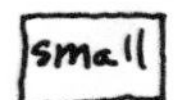

PEOPLE: Whole class, small groups (3-5) and individual

ACTION:

What's in the box?
(Whole class)

Have all the shapes in a closed box. Students take turns pulling one shape out of the box. The students attempt to predict the number of shapes and their attributes before all are drawn from the box. Ask for predictions after each is taken from the box.

Loop activities
(Whole class, then small groups)

One Loop

Spread out one loop. Put a shape inside it. Ask students to describe the shape and choose one of its attributes (color, size, or shape). If, for example, they pick "red," students put all the other shapes that are red inside the loop. The correct place for the remaining shapes is outside the loop.

Pick another shape and repeat the activity.

Two Loops, with labels

Lay out two loops so that they overlap. Choose two Attribute Labels. Place one inside each loop. Pick a shape and ask where it belongs. Some shapes belong outside both loops. Where does the big yellow square belong?

Example A.*

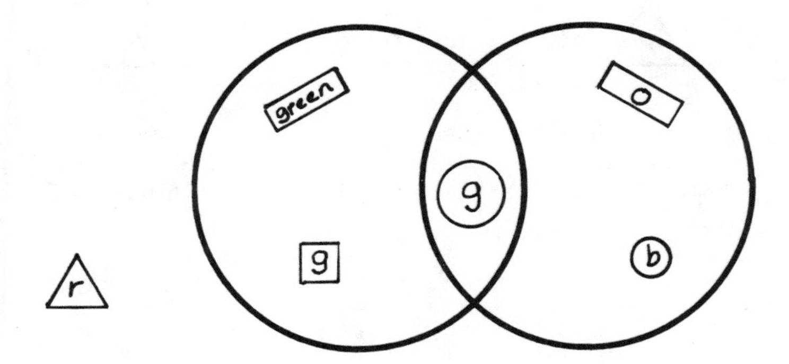

Notice that in this example, the green circles belong in both loops.

*In these examples, the colors will be indicated by "g" for green, "b" for blue, "r" for red, and "y" for yellow.

Two Loops, labels not showing
Again, lay out two overlapping loops. Choose two Attribute Labels and place one *face down* inside each loop.

Start placing shapes in the loops or outside of them until the students think they know what the secret attributes are.

Example B.

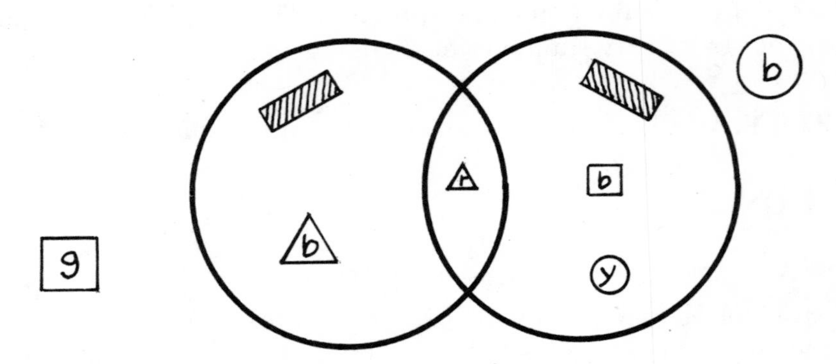

Now have the class divide up into groups of 3-5 students to play 2-loop games among themselves. Each leader spreads out two loops and selects two labels which are placed face down in the loops. A player then selects a shape and places it where she thinks it belongs. The leader tells her whether the shape belongs there or not. Players try to determine the secret attributes, having placed as few shapes as possible.

IF YOU LIKE:
Have students try three loops both with and without secret labels. More Difference Activities:

One Difference
Choose a shape. Have students find all the shapes that differ from it in one way, that is, either by size, color, or shape. For example, if you pick the little red triangle, the following shapes differ from it in only one way:

Original
shape

Same color
Same shape
Different size

Same color
Same size
Different shape

Same size
Same shape
Different color

Two Differences

Choose a shape. Have students find all the shapes that differ from the *large blue square* in two ways:

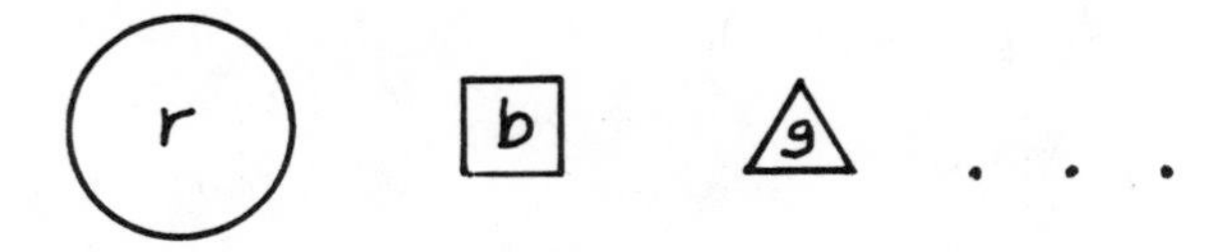

Trains

In groups of 2-4, have students make difference trains.

One Difference Train: Each student in turn selects a shape that is different in one way from the preceding shape.

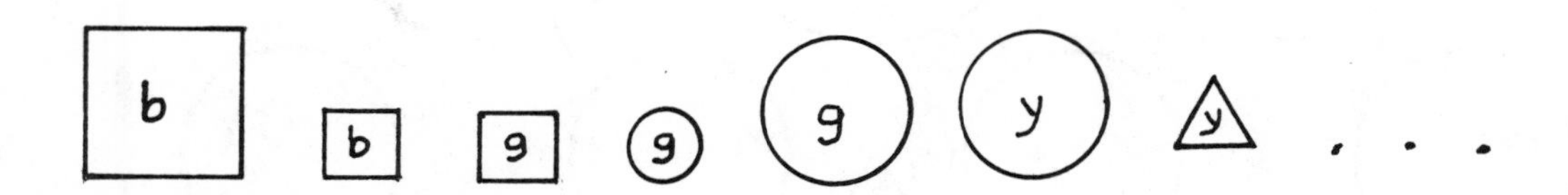

Two Difference Trains: Each shape needs to be different from the preceding one in only two ways.

Attribute Shapes

(ditto sheet)

Spatial Creatures

The defining characteristics of each type of creature are unknown and must be deduced from the examples. This process of delineating possibilities and then organizing and analyzing the data to eliminate possibilities is a powerful problem-solving tool that has applications in many areas of math and science.

- **Attribute identification**
- **Classification**
- **Logical reasoning**
- **Fantasy**

MATERIALS

- **Set of spatial creature cards for each student or pair of students**
- **Crayons or pencils**

PEOPLE: Individuals or pairs

ACTION:

Give each student or pair of students a set of cards. The first card illustrates the process. The task is to find the characteristic or characteristics that distinguish the creatures from one another.

All the creatures on the first row have one or more characteristics in common. None of the creatures on the second row have *all* of these characteristics. The students are challenged to find the creatures on the third row that belong in the first row.

On the first few cards, there is only one defining characteristic. Most cards describe creatures with two unique characteristics. The last few cards feature creatures with three characteristics.

IF YOU LIKE:

Have students make up their own spatial creatures. Suggest that they stick with one characteristic until they are familiar with the process.

SPATIAL CREATURES - Distinguishing Characteristics
A. POLYGONS - st. lines, B. GREEPS - non-straight lines, C. BYCLOPS - 2 eyes, D. PRICKALEEPS - prickly skin and one forked tail, E. CLIPPAWAS - shaded and one eye, F. GLIFFAHUMPHS - 2 eyes and 3 bumps, G. LIMADROOPS - 3 eyes, and 2 appendages: one is 4 hairs and one is arrowed tail, H. HEFFALIPPI -shaded, whiskers, one triangular eye (same height), I. LUMS - one eye and st. lines *or* 3 eyes and curved lines.

SPATIAL CREATURES

(ditto sheet)

As training for future spaceflights, here's your chance to practice identifying strange creatures.
Whether you want to be an anthropologist, zoologist, astronaut or chemist where you will be analysing the old or deciphering the new, these activities may help you sharpen your eyes to subtle difference and help you organize data.
Use the clues given to pick out the distinguishing characteristics of each type of creature. Some have only one distinguishing characteristic, some have two and a few have three.
When you have finished, try making up some of your own to try on your friends.
Can you think of earthly applications which might arise in the above occupations or your life?

A. Pre-flight training card.
These are POLYGONS:

B.

These are GREEPS;

  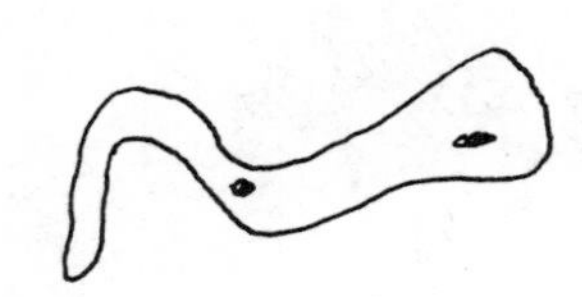 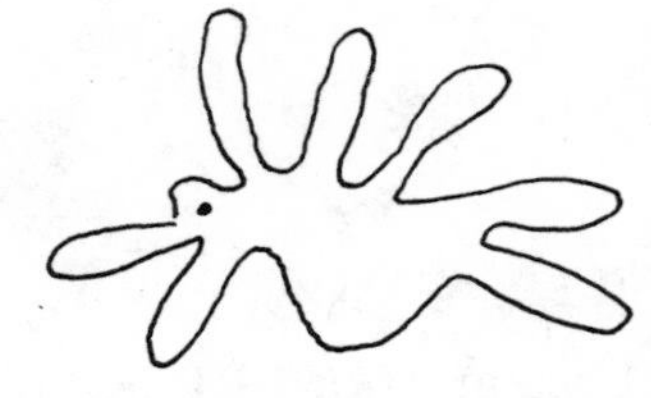

These are not GREEPS:

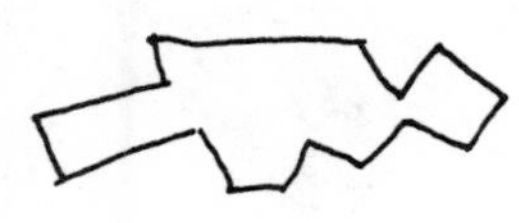   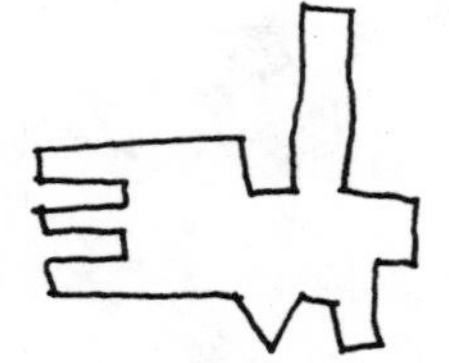

Which of these are GREEPS?

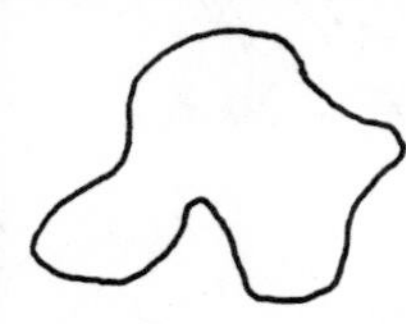  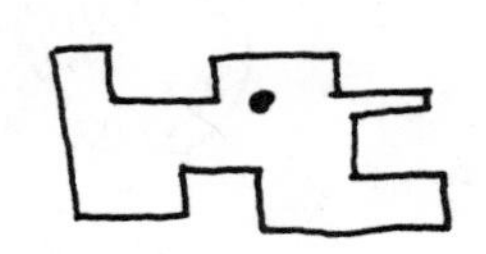  

C.

These are BYCLOPS:

 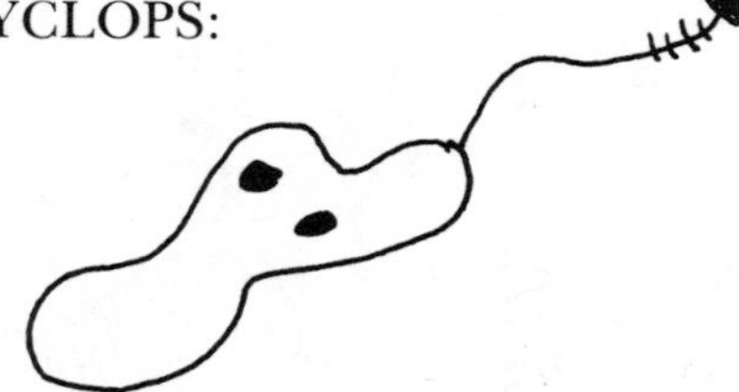  

These are not BYCLOPS:

 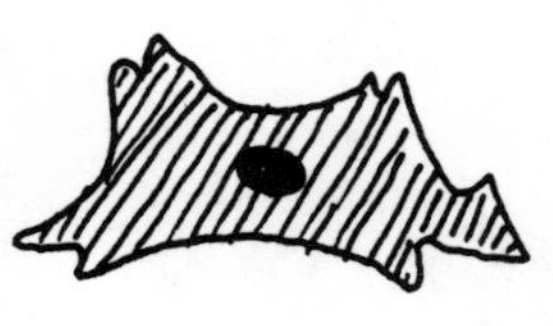  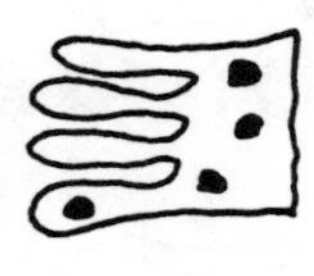

Which of these are BYCLOPS?

D.

These are PRICKALEEPS:

  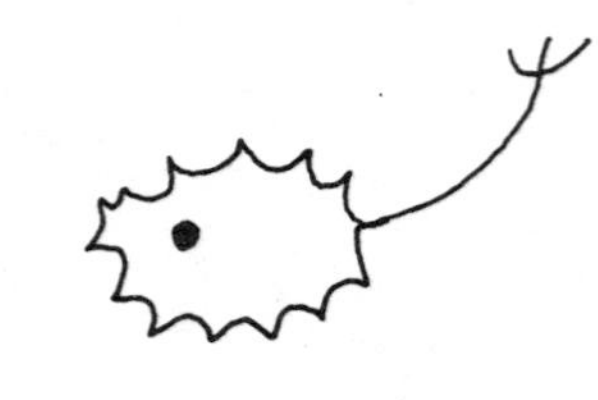 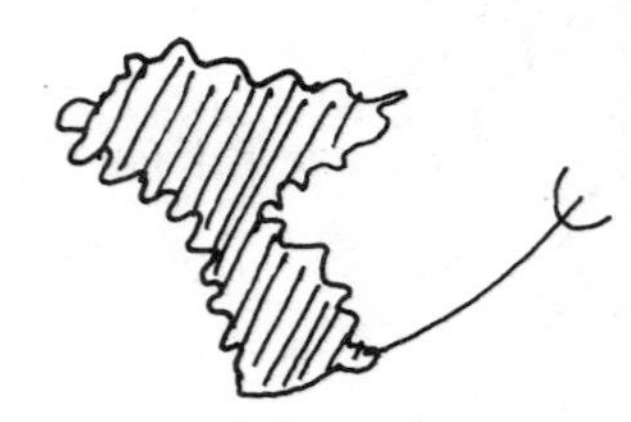

These are not PRICKALEEPS:

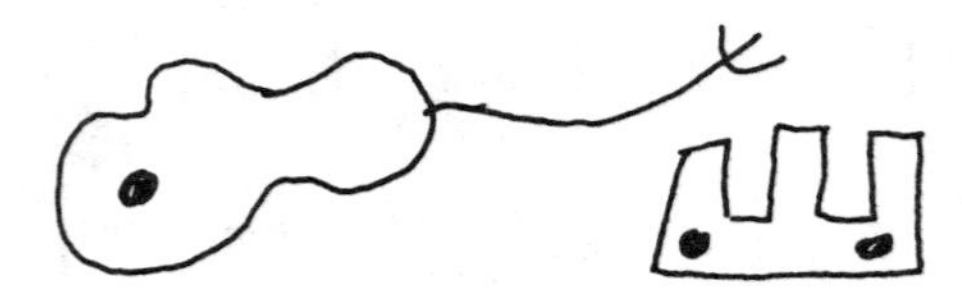  

Which of these are PRICKALEEPS?

E.

These are CLIPPAWAS:

   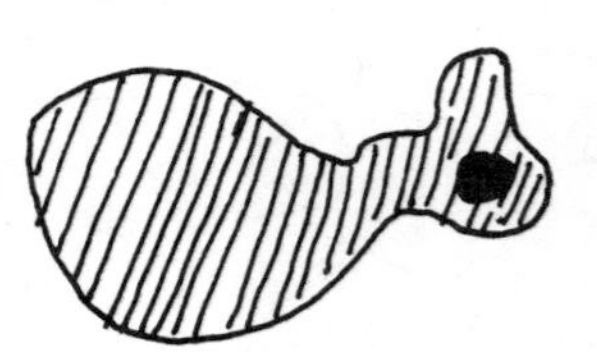 

These are not CLIPPAWAS:

Which of these are CLIPPAWAS?

F.

These are GLIFFAHUMPHS:

These are not GLIFFAHUMPHS:

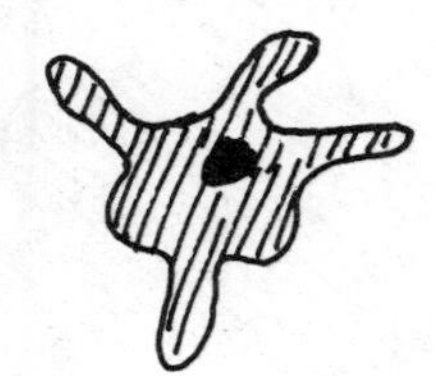

Which of these are GLIFFAHUMPHS?

G.

These are LIMADROOPS:

These are not LIMADROOPS:

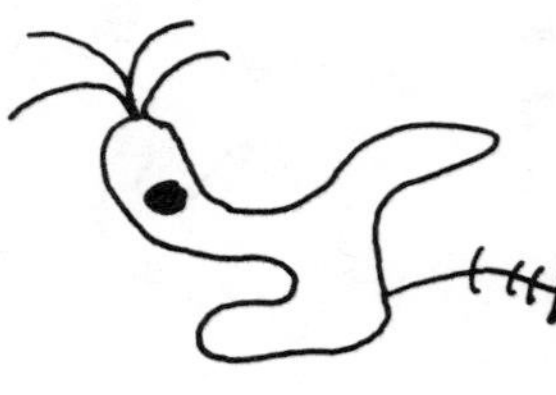

Which of these are LIMADROOPS?

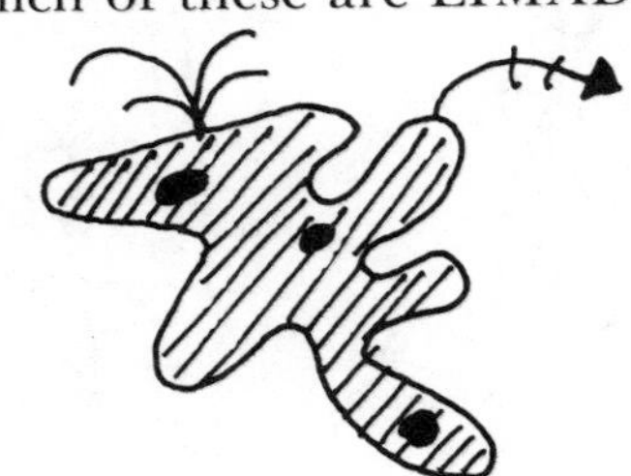

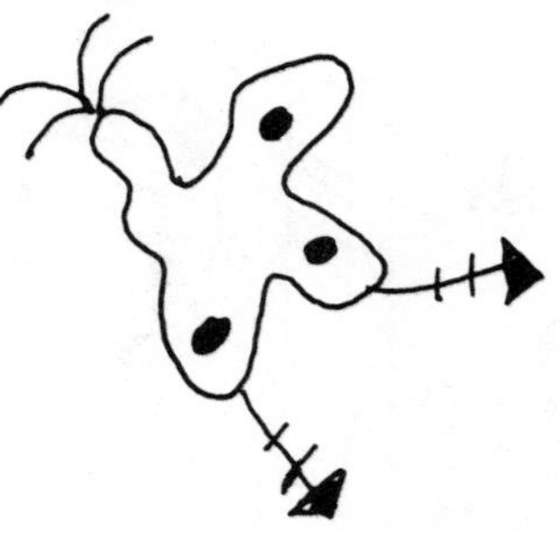

H.

These are HEFFALIPPI:

These are not HEFFALIPPI:

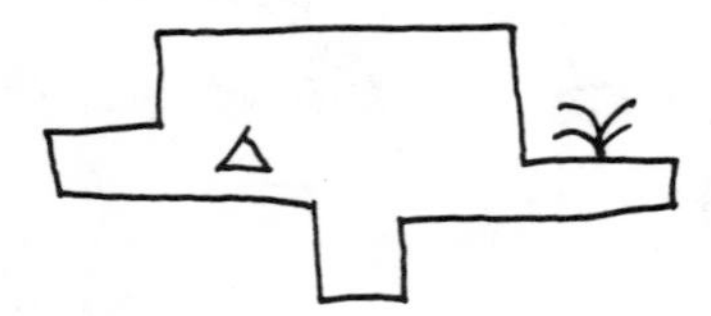

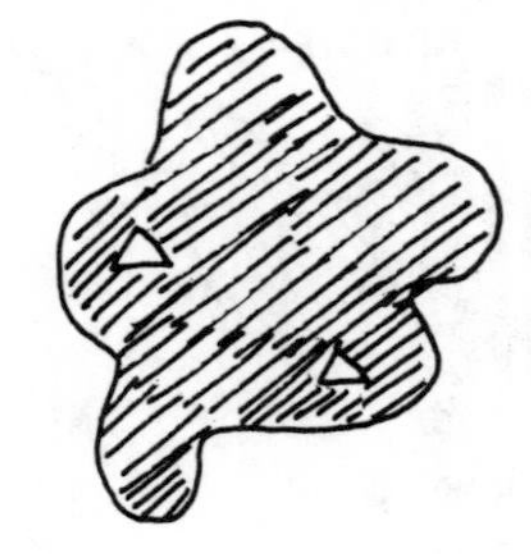

Which of these are HEFFALIPPI?

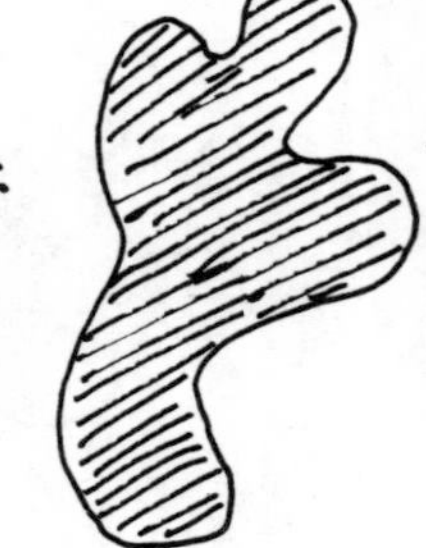

I.

These are LUMS:

These are not LUMS:

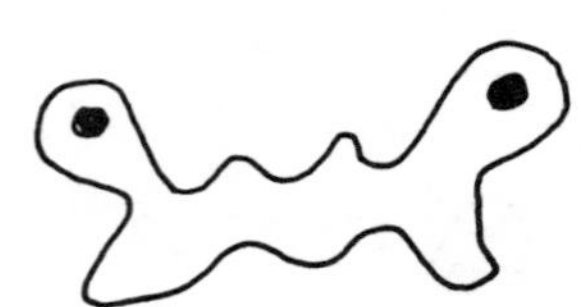

Which of these are LUMS:?

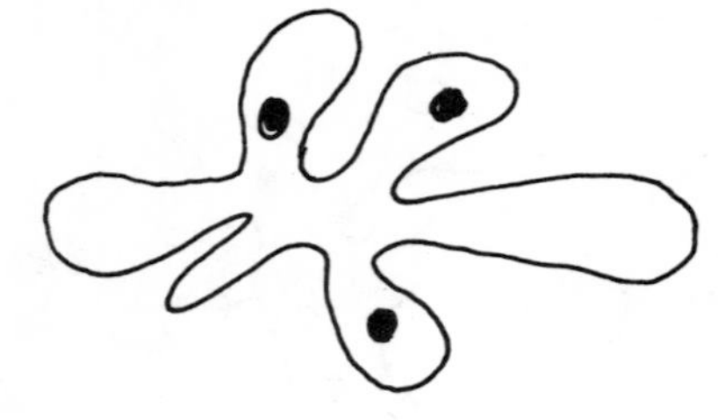
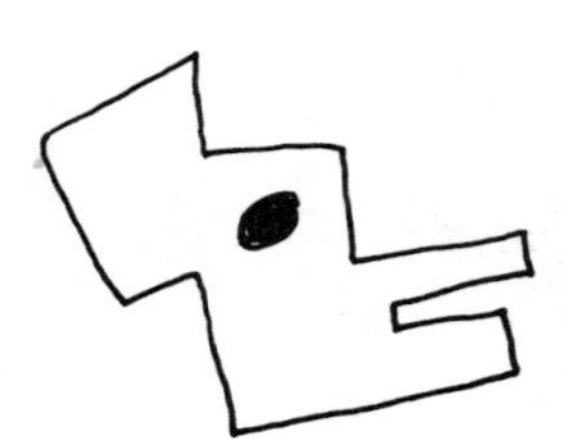

Survey And Graph

- **Formulation of questions**
- **Delineation of range and type of responses**
- **Data collection**
- **Bar graphing**
- **Graph interpretation**

The collection, organization, and interpretation of data are critical to any scientific investigation. Activities of this nature also help students think logically and plan ahead. They learn what questions will yield a lot of information and what questions give virtually no generalizable information.

PEOPLE: Whole class

MATERIALS
- **Pencils**
- **Paper**
- **Graph paper**
- **Butcher paper**
- **Felt pens**

ACTION:

Pick a sample question to demonstrate the activity. For example, "During which month were you born?" Tally the results and draw a bar graph on butcher paper.

Have each student or pair of students generate one or two questions to ask the group.

Allow about 10 minutes for them to circulate, asking their respective questions.

Have them each graph their results and share them with the class.

IF YOU LIKE:

The activity could be extended a week so that the students could survey special groups (or random people).

Day 3

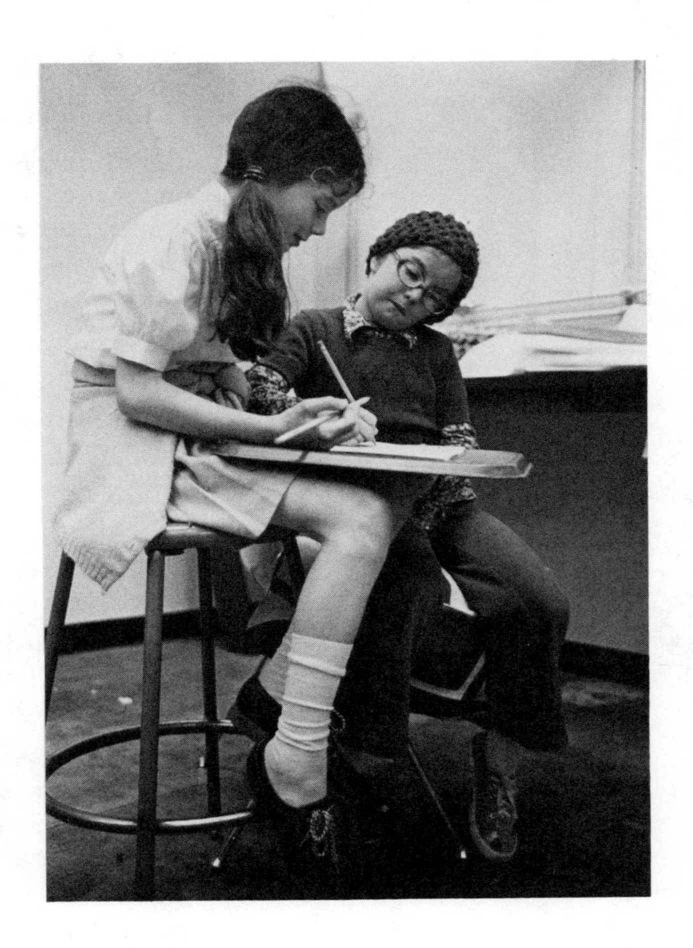

Mystery Stories

- **Group problem solving**
- **Breaking set**
- **Questioning strategies**

MATERIALS
- **None**

This activity stimulates creative thinking. Very often the student must break away from the solution path she has set for herself. Sometimes this can be accomplished by asking questions which lead in another direction entirely.

PEOPLE: Whole class

ACTION:

Tell the students you are going to ask them to solve a mystery. There is one rule — they may ask only questions which have yes or no answers. Any such question, however, is acceptable.

Uplifting Tale I

Tell the following story: I live on the 15th floor of an apartment building. My friend, Emily, comes to visit me every Saturday. Last year she would always ride the elevator to the 7th floor and walk up the rest of the way. This year she rides the elevator to the 9th floor and then walks the rest of the way. Why does she do this? (Additional clue: On rainy days she rides the elevator all the way up to the 15th floor).

Solution: Emily is a child. Last year she could only reach up to the button for the 7th floor. This year she's grown and can reach up to the button for the 9th floor. On a rainy day she brings her umbrella and can reach the button for the 15th floor with the tip of her umbrella.

If the group runs out of questions, give hints about which facts are important. For example:

- What is different about the buttons for 7th and 9th floors?
- How is Emily different this year from last year?

Uplifting Tale II

Tell the following story: I live on the 20th floor of an apartment building. My friend, Emile, comes to visit me every Saturday. On sunny days she rides the elevator to the 10th floor and walks the rest of the way up. On rainy days she takes the elevator all the way to the 20th floor. Rain or shine, however, when Emile leaves, she always rides the elevator from the 20th floor down to the ground.

Solution: Emile is a midget and can reach only as high as the 10th floor button. On rainy days she uses her umbrella to push the 20th floor button. When she rides down in the elevator, she has no trouble reaching the first floor button.

IF YOU LIKE:

Have small groups make up mystery stories. Discuss what makes the mysteries difficult to solve. What are examples of other facts that might surprise us?

Pentro

As in many other strategy games, in "Pentro" it is helpful to plan ahead and to consider the possible moves of the opponent. This will encourage students to think about several variables at one time.

• **Strategy**

MATERIALS
- **6 x 6 grid**
- **Counters in 2 colors**

PEOPLE: Two

ACTION:

The object of the game is for a player to get five counters of her color in a row either horizontally, vertically, or diagonally.

Players take turns placing a counter on any vacant intersection (a place where two lines come together).

Strategy: After several games have been played, ask the students to think about the following questions: Would you rather play first or second? What configurations of counters would ensure your winning after your next turn?

IF YOU LIKE:
- Try this game so that 3, 4, or 6-in-a-row wins. In which cases is a winner quickly determined? In which cases is a winner unlikely to be determined?
- Play on a 10 x 10 grid. Change the goal so that the person who gets the most 5-in-a-rows wins. The game ends when neither player can place a counter to her advantage.
- Play on a larger grid with 3 or 4 players. When does cooperation become necessary?

The Man in the Pit

- **Open-ended problem solving**
- **Creative thinking**

MATERIALS
- **Pencils**
- **Paper**

When presenting this problem, if teachers emphasize the fact that there is no right answer and that the purpose is to find any solution that works, students will have the opportunity to use their imagination with little risk or fear of judgment. An open-ended problem should be versatile enough to allow each child to use her knowledge, however limited or extensive.

PEOPLE: Individuals or small groups (3-5)

ACTION:

Present the problem:
A man, wearing only his hiking shorts, got lost during a summer hike and fell into a pit. The pit was circular. It was about 30 feet across and 20 feet deep. The man found that all the walls of the pit were hard stone, very smooth, and straight up and down, so there were no hand-holds or foot-holds for climbing. The bottom of the pit was also hard as stone. A stream of water kept flowing over the edge of the pit and ran down one wall. Even behind the waterfall the wall was smooth and straight up and down. After reaching the bottom of the pit, the water then disappeared into a small hole in the floor of the pit. The pit was completely empty except for three things:

1. Exactly in the middle of the pit was a tree growing straight up.
2. Near the tree was a loose flat rock.
3. Two boards were also near the tree.

Figure out how the man escaped from the pit.

Have each individual or group share its solution.

IF YOU LIKE:

Other problems and questions for open-ended solutions:

- Have you ever seen a ship inside a bottle? How do you think it got there?
- You are locked in a room. It's raining outside and there's a leak in the middle of the ceiling. There is nothing in the room except an old dirty coffee cup and a hole in the middle of the room. The hole is a little bit larger in diameter than a ping pong ball and longer and narrower than your arm. If you can get the ping pong ball out of the hole, you can come out of the room.

Pentominos

Students first find all possible ways to form five squares into a pentomino. This requires some systematic means of identifying all the different formations. When the students are then asked to predict which shapes can be folded into boxes, they need to use spatial skills for mentally visualizing flips and rotations. They then have an opportunity to build models to test their ideas and predictions.

- **Spatial visualization**
- **Geometry**
- **Patterns**
- **Problem solving**
- **Constructions**
- **Combinatorics**

MATERIALS
- **2 cm graph paper**
- **5 large squares for each group**

PEOPLE: Individual and small groups (2-3)

ACTION:

Tell the group that they are going to try to find out how many pentominos forms there are. A pentomino contains 5 squares and each square must share a full side with at least one other square.

This is a pentomino:

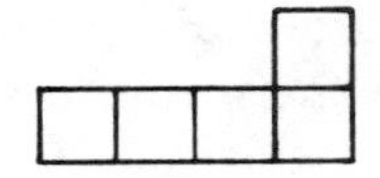

This is not:

Divide the class into groups of 2-3. Have each group form as many pentominos as they can with the five large squares and then draw them on graph paper.

Challenge each group to find as many different pentominos as they can.

The groups then share and compare results with everyone else until different pentominos have been identified.

Next, the students attempt to determine which pentominos will fold into boxes without tops. Ask them also to mark the square that forms the bottom of the box. The students can then cut out the pentominos and test their predictions

IF YOU LIKE:

Explore the number of shapes that can be made using 6 squares. Identify those that will fold into a cube.

Function Machines

- **Functions**
- **Finding patterns**
- **Arithmetic practice**

MATERIALS
- **Blackboard**
- **Chalk**

The willingness to guess, even before there is any evidence of the correct solution to a problem, is an important prerequisite for successful problem solving. Students are encouraged to predict the function machine's output and search for a pattern that will describe the function. Function machines are also useful for expanding a student's understanding of the number system. Described below is a sample classroom introduction to function machines.

PEOPLE: Whole class or small groups

ACTION:

A function machine takes in numbers or other data (input), performs a fixed rule, and gives an "output". Each input results in only one output. These are recorded on the board until students guess the rule.

IN	OUT
2	5
6	9
4	?

Select a secret rule (for example, "add 3"). Students take turns giving input numbers. Before giving the output, ask the students to predict the outcome. Record these guesses on the board.

When a student thinks she knows the rule, she does not describe the rule, but gives an input and predicts the output. She might also predict the output for the numbers others are putting in the machine.

When at least half of the people in the class believe they know what the rule is, ask them to describe their rule.

Continue using simple rules until the function machine is quite familiar to all students. Gradually introduce more difficult, as well as nonnumeric, functions. The following are some examples and suggestions.

Numerical Function Machines:
- add 5 to the input number.
- multiply the input number by 3.

Subtraction rules lead to negative numbers, and consequently are useful for expanding the number system. With younger students, the machine is said to "blow up" if an input number results in a negative output. Similarly, rules involving division can provide an introduction to fractions and decimals for older students. Otherwise, non-integer outputs could cause the machine to blow up. For example, if the rule is to divide the input number by 2, the machine would only process even numbers and blow up for all odd numbers.

Some more complicated functions are:
- multiply the input number by 2 and add 3.
- multiply the input number by 3 and subtract 1.
- If the input number is even, 0 comes out, while if the input number is odd, 1 comes out.
- divide the input number by 5; the output is the remainder.

Nonnumerical Function Machines:

Rules that accept words as input are especially useful for illustrating that, even after several input-output combinations have been seen, several different rules might apply. Each additional guess gives another clue and eliminates possibilities.

Some sample rules are:

- input words, output the second vowel in the word.
- input polygons, output the number of vertices.
- input letters of the alphabet, output 0 if the letter forms an open curve (such as the letter "C") and 1 if it contains at least one closed region (such as the letter "A").

IF YOU LIKE:

Once students have had experience guessing outputs and discovering patterns for rules, strategies can be discussed. Ask the students what clues were helpful. Have them analyze possible rules that are eliminated at each step.

Fictionary

- **Creative thinking**
- **Dramatic arts**
- **Language arts**
- **Vocabulary**

MATERIALS
- **Dictionary**
- **Paper and pencil for everyone**

Often what is most difficult about generating new and inventive ideas is simply overcoming the fear of being judged. Many students lack the opportunity to be creative in a non-threatening environment. This activity and the career skits which follow provide two different media for creative expression.

PEOPLE: 3-6

ACTION:

The first leader goes through the dictionary and finds a word that nobody in the group knows. She then writes down the first dictionary definition for this word.

Each person in the group then secretly writes down her own definition of the word, trying to make her definition sound as much like a dictionary definition as possible.

All the definitions are then folded up and passed to the leader. She then reads all of them, including the real definition, and assigns a number to each definition.

On the count of 3, each person indicates with her fingers which definition she believes is the real one.

SCORING:

Anyone who guesses the correct definition scores 15 points. Each player gets 10 points for anyone who guesses her definition. The leader scores 25 points if no one guesses the correct definition.

Career Skits

- **Career awareness**
- **Creative expression**
- **Organization**
- **Dramatic arts**

MATERIALS
- **Cards with job titles on them (see below for suggested jobs)**

One of the goals of the career component of this class is to increase awareness and understanding of different jobs. Students might discuss what preparation is required and what value each places on the different jobs.

PEOPLE: Groups of 3 and whole class

ACTION:
In groups of three, pick three cards with job titles.
Prepare a skit involving the three occupations (approximately 3 minutes long).
Present the skit and see if everyone else can identify the occupations. Students may not use the name of the profession in their skits.

IF YOU LIKE:
Give each group a scenario as well as their occupations or allow them to pick the occupations from the list generated by the class.

JOB LIST FOR JOB CARDS:

Bookbinder	Bank teller	Police officer
Photographer	Lawyer	Teacher
Blacksmith	Judge	Librarian
Truckdriver	College professor	Auto mechanic
Cashier	Chef	Model
Postal clerk	Waiter	Carpenter
Secretary	Barber	Electrician
Computer programmer	Firefighter	Plumber
	Engineer	Locksmith
Roofer	Geologist	Dentist
Pilot	Meteorologist	Optometrist
Flight attendant	Oceanographer	Veterinarian
Busdriver	Mathematician	Nurse
Architect	Astronomer	Doctor
Jeweler	Ballplayer	Dancer

Day 4

Mystery Person

- **Estimation of physical attributes**
- **Observation of detail**
- **Accuracy**
- **Proportionality in art**

MATERIALS
- **Costume for mystery person**
- **Bag of toy frogs (or some other small interesting objects)**
- **Butcher paper for drawing composite picture**

This activity illustrates how superficial observations can be. It is easy to overlook details of everyday objects and events. Yet in science and mathematics, as in most other fields, these fine details might be critical for solving a problem.

PEOPLE: Whole group

ACTION:

The whole group sits in a circle and the leader begins a mystery story.

A mystery person in unusual garb enters the room. She disrupts the class and attracts attention by dropping a container of toy frogs on the floor. The toy frogs are quickly recovered and the visitor leaves the room. The group is then asked to reconstruct an accurate picture of the mystery person. First, the group attempts to recall as many facts as possible about the visitor. This should eventually include an average estimate of height, weight, and age.

Once the students reach a consensus about the description, the class attempts to draw a facsimile of the visitor on butcher paper. Assign each pair a section of the body to draw.

Once the description is complete, the mystery person reappears. The students compare their estimates and descriptions with the actual measurements and facts.

IF YOU LIKE:

- To test the students' observation of detail, present the following questions and tasks:

  Does your bedroom door open into the room?

  Which way does the water in the bathtub turn when it goes down the drain.

  Can you draw a picture of a phone dial?

- *Graphing the ghost,* makes a picture of the "class average." Divide the group into pairs. Give each pair an attribute on which to collect data, such as height, color of hair. Each pair records the response of each person in the group to their particular question. Averages are determined and tallies made.
  The group then makes a composite drawing of the "average" student in the class.

- Take the group for a short walk. Each person is asked to make one secret observation on the way. Upon return, each one asks the rest her question, such as "What color is the flagpole?" or "How many doors on the left side of the hall?"

Geoblock Activities

- **Spatial visualization**
- **Geometric constructions**
- **Communication**
- **Cooperation**

MATERIALS
- **Geoblocks or kindergarten blocks***

These activities stimulate exploration of properties of various geometric shapes. In the process, spatial visualization skills are exercised as one person listens to a verbal description and attempts to replicate a structure from the visual image she forms. The vocabulary needed to distinguish between the different shapes and sizes can grow naturally from communication activities such as "Describe a block" and "Build a structure." Students recognize the importance of precision, common experience, and vocabulary when trying to communicate.

PEOPLE: Small groups or pairs

ACTION:

Find a Block

Each person works with a partner and a small set of blocks. One person closes her eyes and the other person gives her a block. She may feel it until she knows it and then it is returned to the pile. When she opens her eyes, she tries to find the block among the other blocks.

Describe a Block

Divide the class into small groups. One person thinks of a particular block. The rest of the group tries to guess which block has been chosen by asking questions which can be answered by "yes" or "no," such as "Does it have any square sides?" The questioning continues until the group successfully identifies the chosen block.

Build a Structure

The students work in pairs, with each having the same 4 blocks. Place a barrier between each pair. One person builds a structure with her blocks. She then tells her partner how to build the same structure without watching.

When the description and construction of the structure is complete, compare structures. Are they the same? If not, where was the misunderstanding? How could it have been avoided?

IF YOU LIKE:

Have everyone switch roles.

*These activities can be done with any blocks. However, the variety of shapes in the Geoblock set add interest and complexity to each activity.

Topological Rope Puzzles

- Spatial visualization
- Topology concepts
- Closed and open curves
- Knots
- Topological equivalence

MATERIALS
- Handcuffs made of rope or string
- Rope rings (3" diameter)

These puzzles provide a challenging setting for visual thinking and a different approach to a baffling problem. Students learn new techniques for problem solving like building and solving simple or equivalent models of the problem.

PEOPLE: Two

ACTION:

Handcuffs may be made of light rope or string. The loop for each wrist may be made with a slip knot or may be tied loosely to each person's wrists.

Puzzle 1: Handcuff students together in pairs (see diagram). The two ropes should criss-cross as shown, so that the two people are fastened together.

The problem is for the two to separate themselves from each other without removing the rope from their wrists.

Let the pairs try any approach they want for about five minutes or until they become frustrated, then introduce each pair to the simpler versions of the problem.

Puzzle 2: One student puts one set of handcuffs over one arm with a ring on the rope as pictured:

Object: Separate the ring from the rope and the arm.

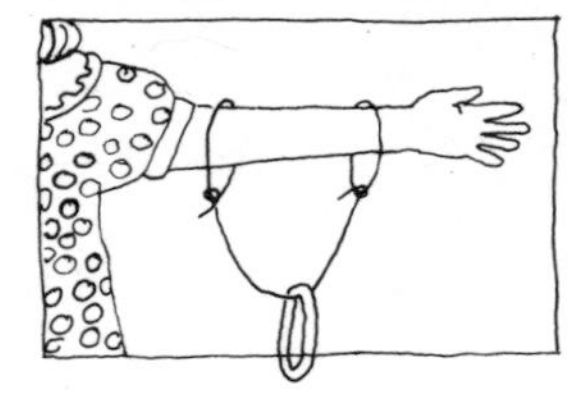

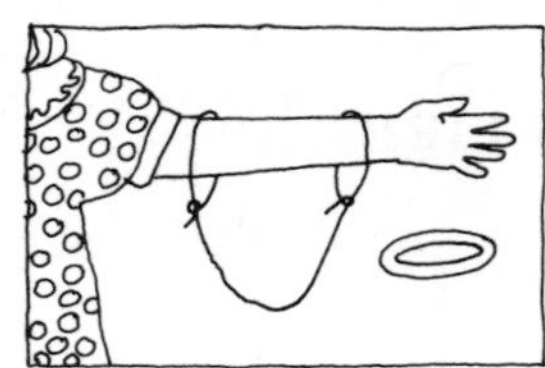

Puzzle 3: One students put on one set of handcuffs. Again the ring is on the connecting rope.

Object: Remove the ring from the connecting rope

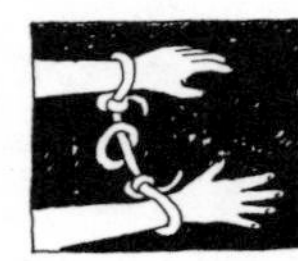

Puzzle 4: Now the students go back to the original handcuff puzzle.

Solution to Puzzle 2:

Answers to these topological puzzles are not appropriate, since discovery is important to internalizing the concepts and since several topologically correct answers are possible for most of the puzzles. However, to help with frustration, here are some basic steps for Puzzle 2, which can be applied to all of the other puzzles shown with minor variations.

1. Put Loop A from its original position over one hand.
2. Then slide the "hand-cuff" loop through Loop A.
3. Insert Loop A through Loop B

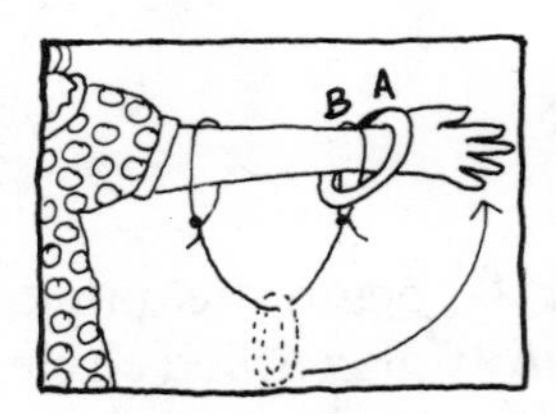

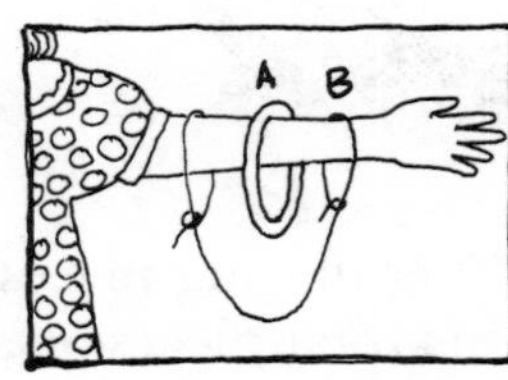

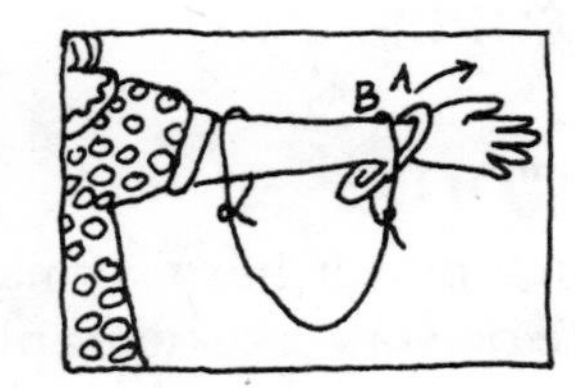

4. Slip Loop A off the hand!

IF YOU LIKE:

A set of wood and rope puzzles that model the handcuff puzzle can be constructed as illustrated below.

I. One pole handcuff puzzle

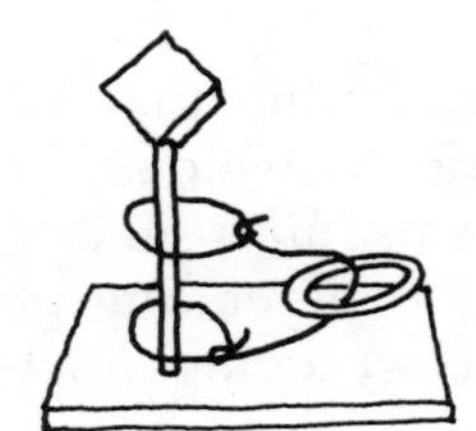

Problem:
Remove the loop

II. Two pole handcuff puzzle

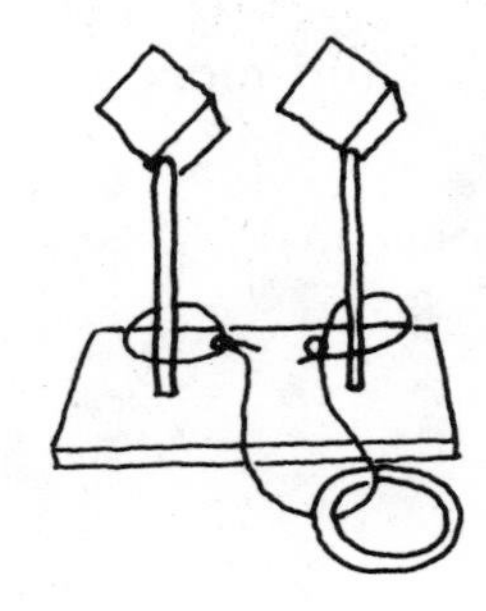

Problem:
Remove the loop.

III. Four pole puzzle

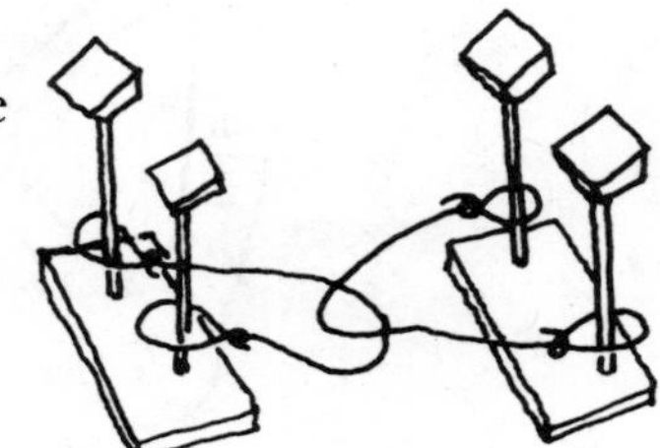

Challenge:
Separate the two parts of the puzzle.

Horseshoe Game*

- **Network**
- **Strategy**

MATERIALS
- **Horseshoe network**
- **Two playing pieces for each player**

A *network* is a collection of points (or locations) and lines (or paths) connecting pairs of points. Horseshoe is played on a network with five locations and seven paths. This network game provides a simple example of a game with only one possible ending position. As such, it can be analyzed by starting at the ending position and working backwards. This strategy is called backtracking.

PEOPLE: Two

ACTION:

Each player has two matching playing pieces. To begin the game, the players take turns placing their pieces on any unoccupied circle on the network. After all the pieces are on the board, play continues by moving to adjacent circles. No jumps are allowed. An opponent's pieces are captured and the game ends when the opponent cannot move any of her pieces.

In the example below, it is B's turn. Since the only empty circle is not adjacent to either of B's pieces, B cannot move. Hence the game is over and A is the winner. This is the ending position for the game. A player's pieces could be cornered on the left side of the board rather than the right side; these are the only ways the game can end.

It may take players quite a while to finish a game since once the end position is known, it can be avoided. The students may want to discuss why "capture" seems impossible.

Working backward from the ending position, it is possible to enumerate all the possible preceding moves and figure out how to avoid being cornered.

Players may want to decide that after a certain number of moves the game is a draw.

The next game is similar and more satisfying.

*A Network Game by Dennis Sullivan.

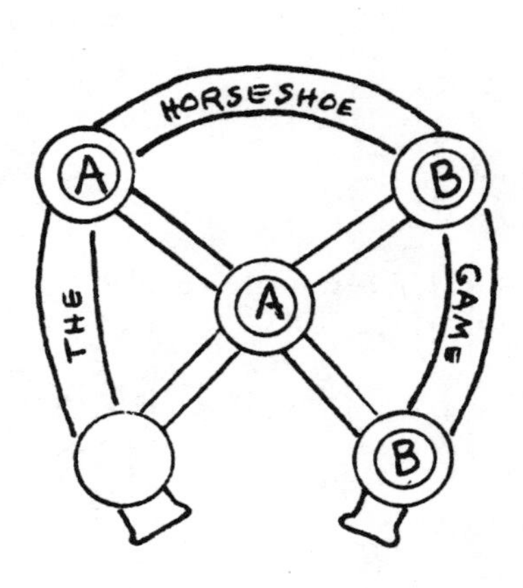

THE HORSESHOE GAME (ditto sheet)

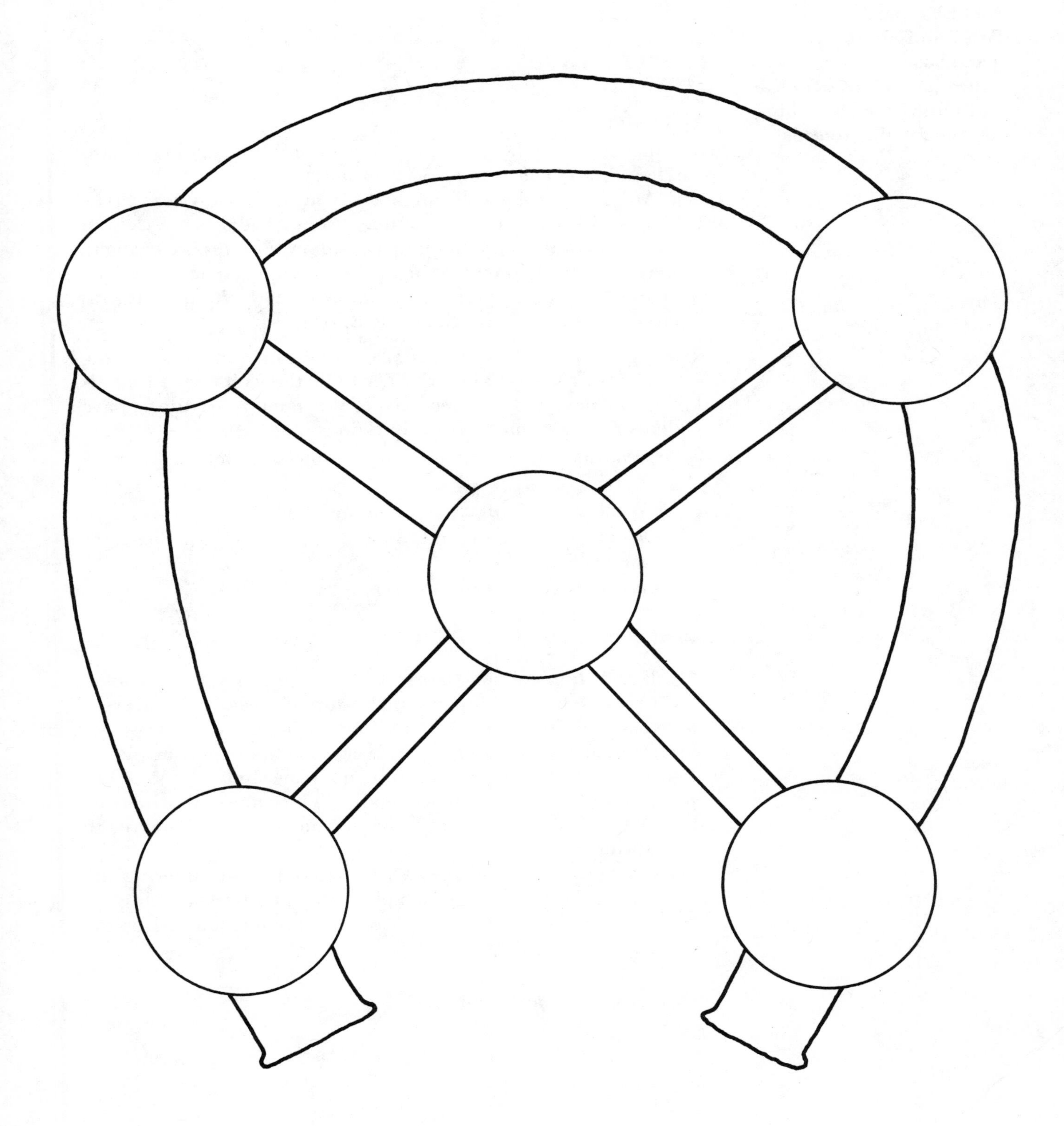

Wolf And Hounds*

- **Network**
- **Strategy**

MATERIALS
- **Wolf and Hounds gameboard**
- **One playing piece for the wolf and three matching pieces for the hounds**

This network game provides an opportunity for thinking ahead and backtracking in attempts to find winning strategies. One challenge might be to figure out if it is better to be the wolf or the hounds.

PEOPLE: Two

ACTION:

One player moves the three hounds who start on circles A, B, and C. The other player is the wolf who starts on W. The wolf moves first. At each turn the wolf can move to any empty adjacent circle. The hounds also move, one at a time, to empty adjacent circles but may move forward only. No jumps are allowed. They are trying to corner the wolf so that it does not have an empty adjacent circle to move into. The wolf is trying to get away from the hounds. It wins by landing on any of the circles A, B, or C.

Strategy: Encourage students to play 3 or 4 games in a row. Let the loser of each round decide whether to be the wolf or the hounds during the next game. After playing the game a few times, have students discuss some of these questions:

- Who has the best chance of winning, the wolf or hounds? Why?
- What would happen if the hounds move first?
- Where are the best (and worst) places on the gameboard for the wolf?
- What moves should the hounds avoid?

IF YOU LIKE:

"Wolf and Hounds" and "Horseshoe" are only two of many possible network games. Suggest that students make up different networks consisting of points and connecting lines. These can be used as gameboards. Give each player one or more playing pieces. Players alternate turns. At each turn a player moves one of her pieces along a line to an adjacent point. The game ends when one player is cornered and can no longer move. The last player to move wins.

In general these games can be analyzed by considering the number of possible moves each player has and by examining the number of turns needed to get from one point to another on the gameboard.

---

*A Network Game by Dennis Sullivan

WOLF and HOUNDS

(ditto sheet)

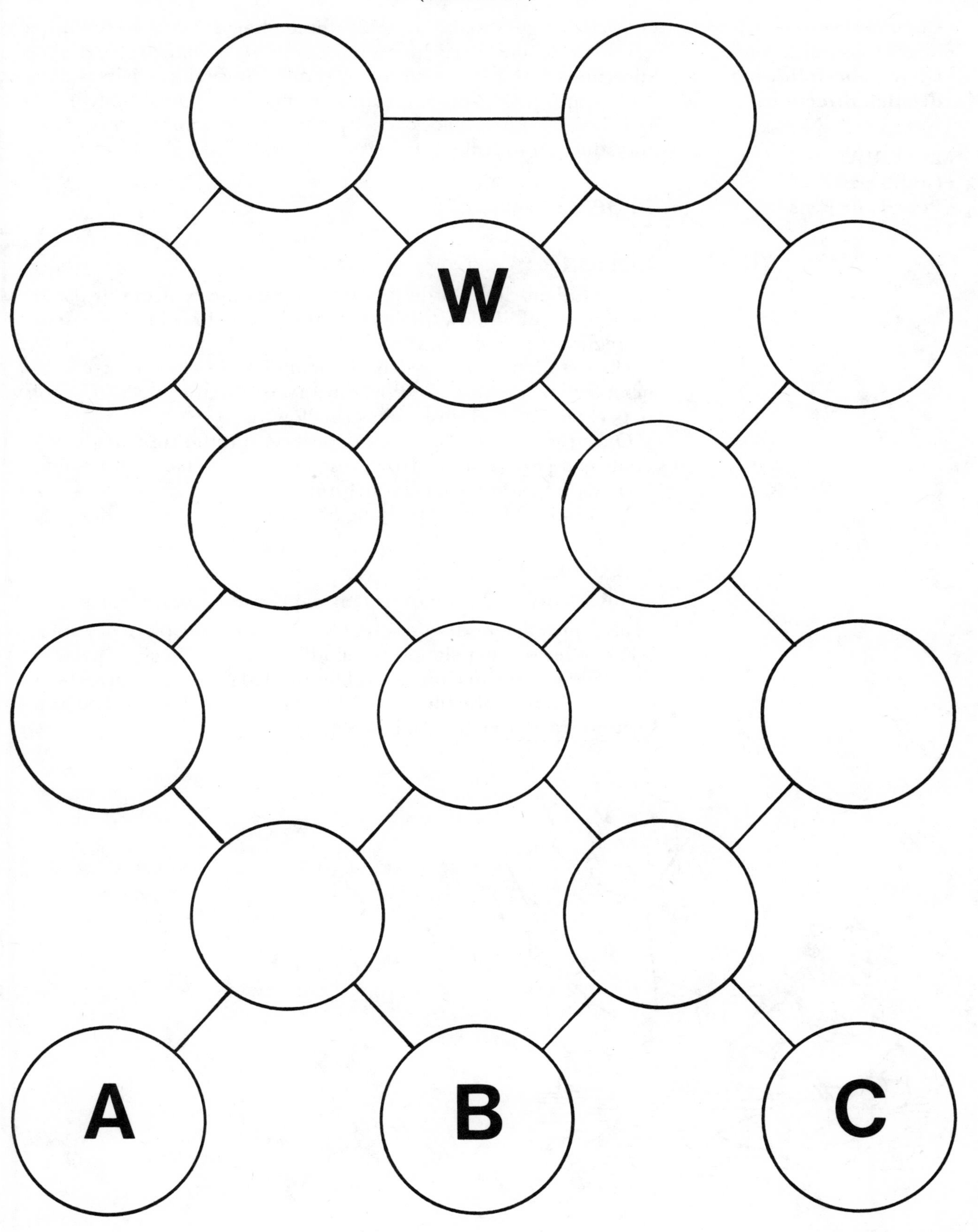

Double Design

- **Communication**
- **Spatial visualization**
- **Giving and following detailed directions**

MATERIALS
- **Graph paper**
- **Pencils or pens**

The challenge is to duplicate an unseen line drawing by following verbal directions. Participants must not only be able to give precise directions, but they must listen carefully. Since the activity is done with graph paper, the directions are necessarily quantitative, unlike those in the Geoblock activity. This activity can also provide an introduction to scale drawings.

PEOPLE: Groups of 2-4

ACTION:

Each student draws a design on her own piece of graph paper. Designs should follow the lines on the paper. Encourage students to begin with simple designs.

Have students pair up. One student describes her picture while her partner attempts to reproduce it on a blank sheet of graph paper. Only verbal directions are allowed.

Once the drawing has been described and the copy made, students compare designs. How close are they? What additional information would have been helpful?

Now have the students change roles.

IF YOU LIKE:

- This activity can be done on blank paper rather than graph paper.
- Have one partner draw a squiggly line on a blank piece of paper. The other partner should be blindfolded. The "seeing" partner places the paper in front of the blindfolded person and gives her a colored pencil. She then gives directions to the blindfolded person to trace over the squiggly line.

Mystery Story

- **Group problem solving**
- **Breaking set**
- **Questioning strategies**

MATERIALS

- **None**

This activity stimulates creative thinking. Very often the student must break away from the solution path she has set for herself. Sometimes this can be accomplished by asking questions which lead in another direction entirely.

PEOPLE: Whole class

ACTION:

Tell the students you are going to ask them to solve a mystery. There is one rule — they may ask only questions which have yes or no answers. Any such question, however, is acceptable.

Woman on the Corner

Tell the following story: A woman is standing on a corner, wearing a mask. Another woman runs up to her, and just before reaching the masked woman, she turns and runs back in the direction from which she came. What's happening?

Solution: The "corner" is home base. The masked woman is a catcher in a baseball game. The woman was running toward home base from third when the catcher caught the ball. So she then turned and ran back to third.

If the group runs out of questions, give hints about which facts are important. For example

- Why do people wear masks?
- Who would stand on a corner — does it have to be the corner of a building?

IF YOU LIKE:

Have small groups make up mystery stories. Discuss what makes the mysteries difficult to solve. What are examples of other facts that might surprise us?

- unusually thin or fat person
- people doing unusual jobs
- words with double meaning.

Day 5

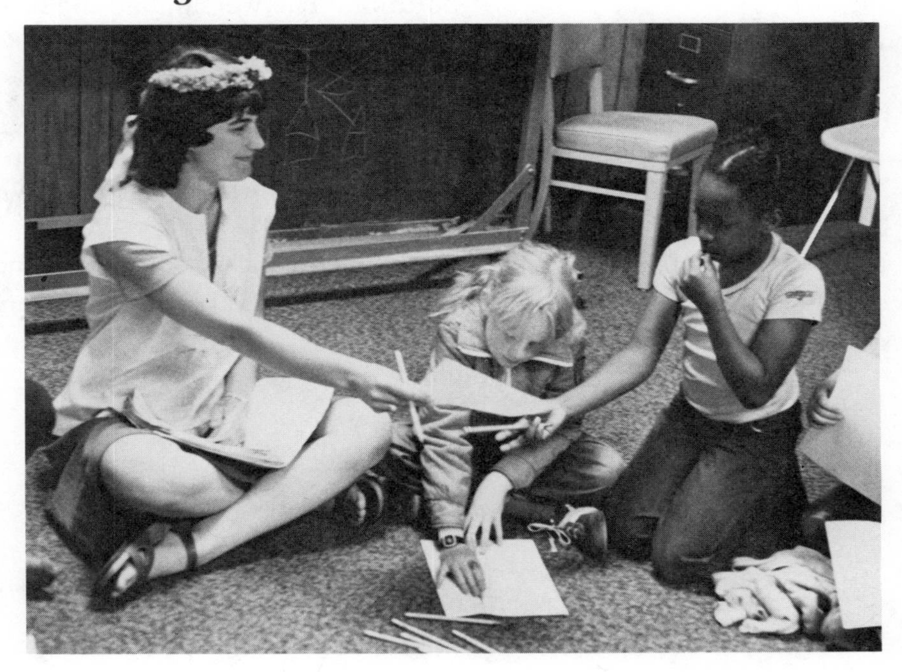

Mystery Story

- **Group problem solving**
- **Breaking set**
- **Questioning strategies**

MATERIALS
- **None**

This activity stimulates creative thinking. Very often the student must break away from the solution path she has set for herself. Sometimes this can be accomplished by asking questions which lead in another direction entirely.

PEOPLE: Whole class

ACTION:

Tell the students you are going to ask them to solve a mystery. There is one rule — they may ask only questions which have yes or no answers. Any such question, however, is acceptable.

Woman in a Restaurant

Tell the following story: A woman walked into a restaurant and asked the owner for a glass of water. The owner pulled out a gun and aimed it at the woman. The woman was very pleased. She thanked the owner and walked out. What happened here?

Solution: The woman had the hiccups and wanted a drink to get rid of them. The owner frightened them out of her.

IF YOU LIKE:

Have small groups make up mystery stories. Discuss what makes the mysteries difficult to solve? What are examples of other facts that might surprise us?

- unusually fat or skinny person.
- words with double meaning, such as, *home* for home plate or a house.

How Many Frogs?

This estimation activity differs from "Guessing the Total Age" in several respects. Here the students are estimating a discrete quantity. Consequently, they can use some well-defined visual estimation techniques that cannot be used when guessing ages. This activity also introduces several ways of recording data.

- **Guessing**
- **Estimation**
- **Data**
- **Exploring range**
- **Frequency**
- **Median**
- **Mean and mode**
- **Graphing**

MATERIALS

- **Jar of less than 100 toy frogs, turtles, gumballs, pennies or other objects**
- **Laminated 100-square (see below)**
- **Laminated grid labelled 10, 20, 30, 40, . . . 100 across bottom (see below)**
- **Counters**
- **Crayons**

PEOPLE: Whole class

ACTION:

Let students take turns holding the jar, shaking it, looking at it from different angles. Each records her guess on the laminated 100-square by finding that number and putting a counter on it. If two or more have the same guess, they can stack their counters on the appropriate number.

Next, each person records her guess on the bar graph by placing an "x" in the appropriate column. For example, a guess of 68 falls between 60 and 70, so she would put an "x" in the column between 60 and 70.

Finally, give each one a handful of objects to count. Write each one's total on the board and have the students find a grand total.

IF YOU LIKE:

- Ask students to use the information just gained to predict the number of frogs in a jar twice this size or 3 times this size.
- Suppose the jar was filled with rice. Find estimates for the number of grains.
- Have students examine the 100-square and the bar graph to determine where the guesses clustered. Ask how many guesses were within 10 of the correct total.
- Ask how the bar graph might compare with a graph of their heights. If appropriate, have students compute the mean and mode. Compare these with the actual number of objects in the jar.

BAR GRAPH

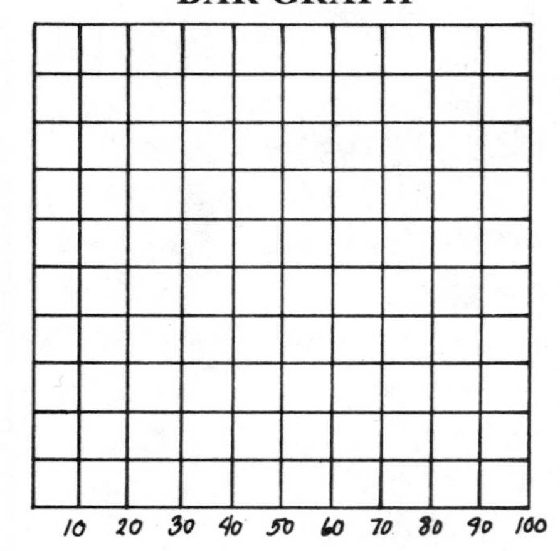

100-SQUARE

1	2	3	4	5	6	7	8	9	10
11	12	13	14	15	16	17	18	19	20
21	22	23	24	25	26	27	28	29	30
31	32	33	34	35	36	37	38	39	40
41	42	43	44	45	46	47	48	49	50
51	52	53	54	55	56	57	58	59	60
61	62	63	64	65	66	67	68	69	70
71	72	73	74	75	76	77	78	79	80
81	82	83	84	85	86	87	88	89	90
91	92	93	94	95	96	97	98	99	100

Toothpick Puzzles

- **Spatial visualization**
- **Problem solving**
- **Breaking set**

MATERIALS:
- **Toothpicks**
- **Toothpick puzzle ditto**
- **Pencils**

Toothpick puzzles abound in mathematics. Usually, a visual image is presented and the problem solver is asked to alter the configuration to produce another by removing or adding a certain number of toothpicks. The breaking of a mind set is frequently necessary. The resulting configuration may look quite different from the original. Students gain confidence by solving simpler puzzles and then tackling more difficult ones.

PEOPLE: Two

ACTION:

Give each pair of students a pile of toothpicks, a ditto, and a pencil. Encourage the students to lay out each configuration with toothpicks and try out many ideas for solving each puzzle. Once a solution is discovered, it can be recorded on the puzzle page. Many of the puzzles have several solutions.

IF YOU LIKE:

Students can make up their own toothpick puzzles. As they attempt to solve the above puzzles, they are likely to come across interesting configurations of their own. Students can create many different puzzles from an initial configuration like number 3 on the ditto.

TOOTHPICK PUZZLES

(ditto sheet)

1. Use 17 toothpicks to construct this figure

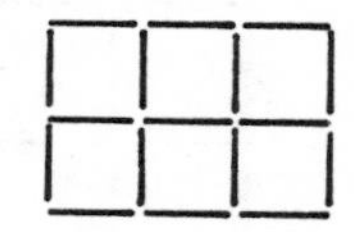

   a. Remove 5 toothpicks and leave 3 squares.
   b. Remove 6 toothpicks and leave 2 squares.

2. Make this figure with 12 toothpicks.

   a. Remove 4 toothpicks and leave 3 triangles.
   b. Move 4 toothpicks and form 3 triangles.

3. With 9 toothpicks, make this figure.

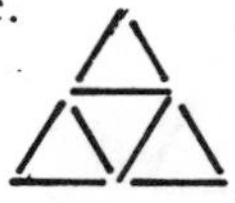

   a. Remove 2 toothpicks and leave 3 triangles.
   b. Remove 3 toothpicks and leave 1 triangle.
   c. Remove 6 toothpicks and get 1 triangle.
   d. Remove 4 toothpicks and get 2 triangles.
   e. Remove 2 toothpicks and get 2 triangles.

4. Use 8 toothpicks and 1 button to form a fish.

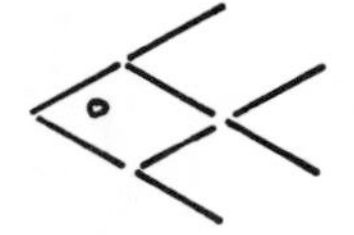

   Move 3 toothpicks and button to make this fish swim the opposite direction.

5. Two farmers have land this shape.

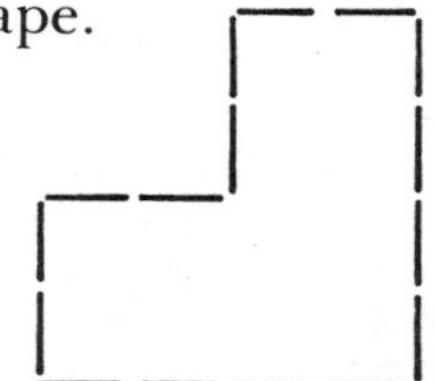

   a. The first farmer wants to divide her land evenly among her three daughters. Add 4 toothpicks to form three parcels of equal size and identical shape.
   b. The second farmer wants to divide her land evenly among her 4 daughters. Use 8 toothpicks to form four parcels of equal size and identical shape.

6. Use 6 toothpicks to form 4 equilateral triangles.

TOOTHPICK PUZZLES — Solutions

There is an "X" on each toothpick to be removed. In most cases there are several possible solutions. Only one is indicated.

1. a.

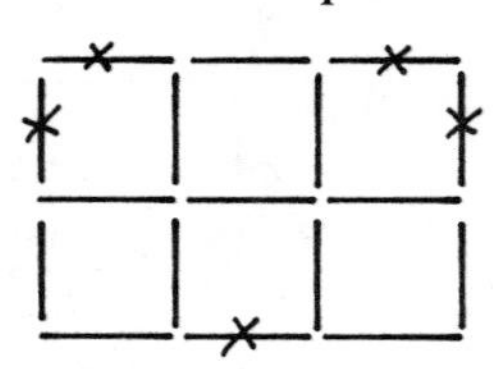

b.

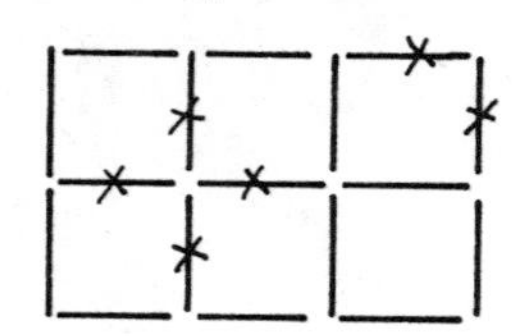

2. a.

b.

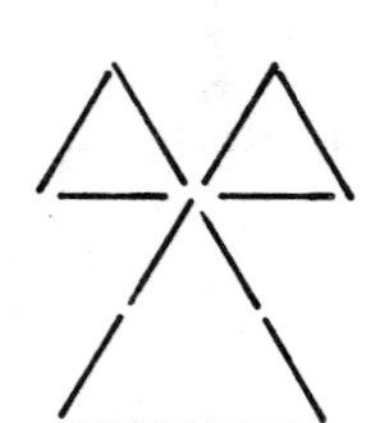

3. a.

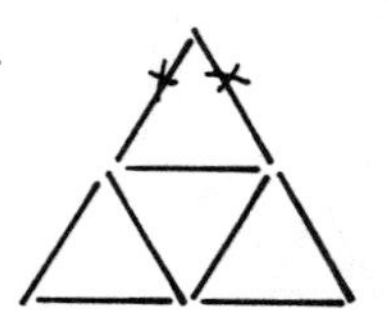

b.

c.

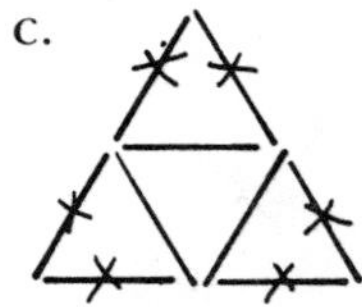

d.

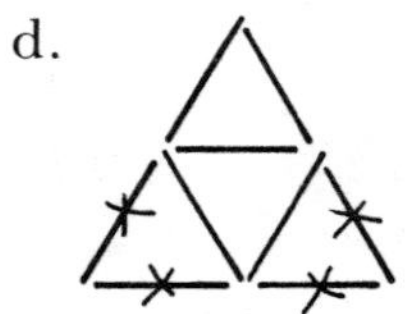

e.

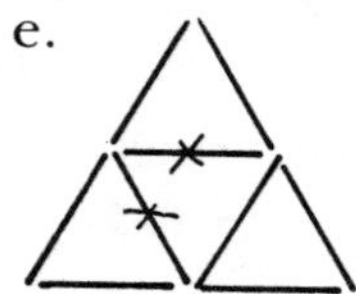

4.

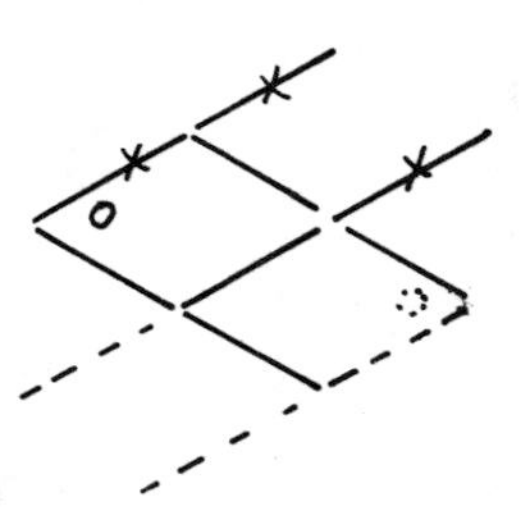

5. a.

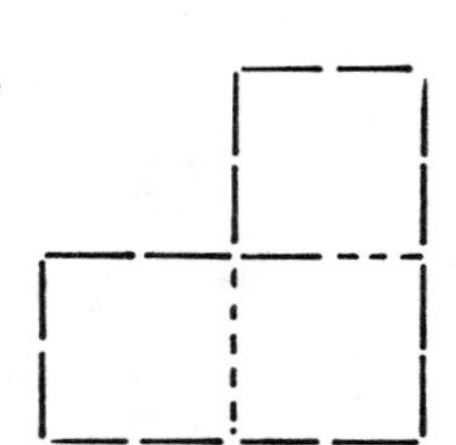

b.

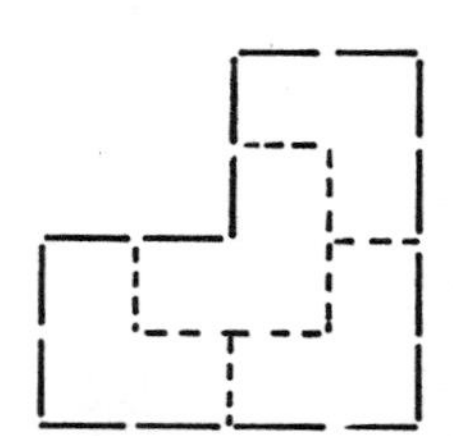

6. Make a 3-dimensional tetrahedron.

Leap Frog

- **Patterns**
- **Strategies**
- **Generalizing**
- **Simplifying a problem**

Students usually solve this puzzle by trial and error. If they begin with a large number of frogs, it will be difficult for them to recognize a pattern in the solution. For this reason, students should have the opportunity to attempt the puzzle with fewer frogs.

PEOPLE: Individuals or pairs

MATERIALS
- **10 green toy frogs**
- **10 red toy frogs**
- **Playing strip (If frogs are not available, use checkers or other markers)**

ACTION:

Line up 2 groups of 5 frogs facing each other on the playing strip, leaving the center square empty. The object is to switch the positions of the red and green frogs. Each frog can only move forward. A move consists of a frog moving forward onto an open square *or* jumping forward over *one* frog onto a vacant space. After the students have struggled with 5 frogs, have them try the puzzle with only 2 frogs on each side, then 3 and 4.

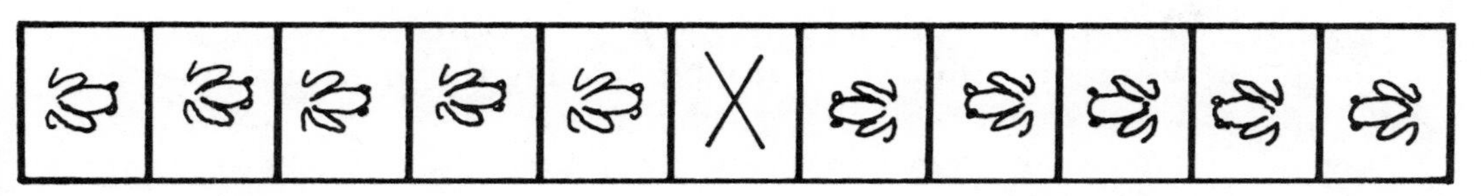

red frogs **green frogs**

IF YOU LIKE:

- Once a solution is found, keep a record of the sequence of moves and total number of moves. Write a "g" when you move a green frog, and write an "r" when you move a red frog. Make a table like this and fill it in.

No. of Frogs on each side	Order of Moves (by color)	Total No. of Moves
1	r-g-r	3
2	r-g-g-r-r-g-g-r	8
3		
4		
5		

When students complete the above chart, ask if they see any patterns. Can they predict the number and sequence of moves needed to interchange 7 frogs of each color?

- Circle the "jump" moves on the chart above. Does this show a pattern?
- The "Tower of Hanoi" puzzle is similar, but a little more difficult to visualize.

Map Colors

- **Problem solving**
- **Spatial visualization**
- **Strategy**
- **Recording data**
- **Art**
- **Geography**

MATERIALS
- **Colored pencils**
- **Felt tip pens or crayons**
- **Paper**
- **Pencils**
- **Blank maps of the U.S.**

In attempting to minimize the number of colors needed to color a map, one must develop a strategy for assigning colors to regions. This process might begin with trial and error and then evolve into systematically examining maps with different numbers of regions. Coloring the map of the U.S. gives students the opportunity to apply this knowledge to a real map. The students can also create their own maps, making them as intricate as they like. Some will discover that an intricate-looking map might in fact be easy to color. By the time students have completed this activity, they may have concluded that only 4 colors are required for any map.

PEOPLE: Individuals or pairs

ACTION:

The group is presented with the following scenario: As a professional map maker, you want to stock as few different colors as possible in order to save money. In order to assure that your customers will have no trouble reading the map, you never color two adjacent regions the same color.

Note: Two regions are not considered adjacent if they have only one point in common.

Present these two sample maps.

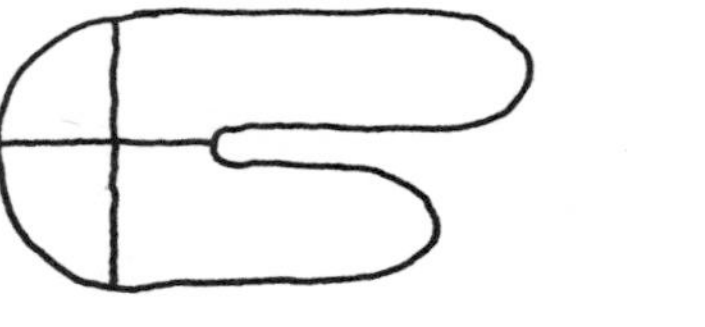

Ask how many different colors are required for each.

In small groups, have students draw maps with 1, 2, 3, 4, 5. . . regions. See how few colors can be used to color each one. Make a chart of colors needed for 1, 2, 3, 4, 5. . . regions. Ask if anyone can draw a 5-region map for which 5 different colors are needed. Pass out the map of the United States. Ask the students to find out how many different colors you have to use.

IF YOU LIKE:

- Everyone draws a map (perhaps not too complicated the first time) and trades with her neighbor. How many colors are needed?
- Ask if anyone can create a map that would require 5 colors.

Find Your Way Home

Coordinates are introduced in terms of directing a person from the origin (0,0) to a house located on the grid. Through finding the location of each other's homes, students discover the coordinate names for points on a grid.

- **Spatial visualization**
- **Introduction to coordinates**

MATERIALS

- **Large grid on chalkboard with coordinates numbered**
- **Small grid for each student**
- **Pencils**

PEOPLE: Groups of 4 or 5

ACTION:

Pretend the grid is a map. Each line is a street and each space is a block.

Draw a house at one of the intersection points of the grid on the blackboard.

Ask students to suggest ways to direct someone standing at the bottom left hand corner (O,O) to the house. After exploring various suggestions, indicate that one way would be to direct the person to walk along the bottom and then up.

For example:

Directions to Meg's house:
3 over and 2 up
(3,2)

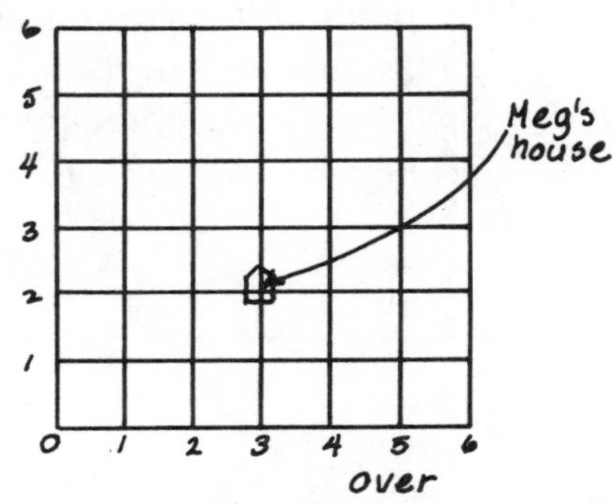

This method gives the coordinate name for that intersection as (3,2).

Divide the students into groups of 4 or 5. Give each person a grid and a pencil.

Each person chooses an intersection to locate her house and marks it on her grid without showing the other players.

Each person in the group takes a turn directing the group to her house from (0,0) by indicating the number of blocks to walk over and then up. Once found, draw the house in at the appropriate intersection and label it with its coordinate name and with the initials of its "owner."

Compare maps. All should be identical.

IF YOU LIKE:

- If this is the first time students have encountered coordinates, have each group work together to produce only one map.
- Have students find distances between houses. Which 2 houses are closest? Which 2 are farthest away?
- Lots of Dots by Dennis Sullivan has related activities. See Bibliography, page107.

Coordinate Dice

- **Strategy**
- **Coordinates**
- **Spatial visualization**

MATERIALS
- **24 tokens each of 2 colors**
- **Numbered grid (see next page)**
- **two different colored dice (one green and one red)**

The coordinate name for each point in the grid is reinforced in this strategy game. Each player tries to get 4 of her tokens in a row vertically, horizontally, or diagonally. At the beginning of the game, chance determines the fall of the dice and placement of the tokens. As the grid fills up and rolls repeat, players can choose the placement of the tokens. The strategy is different from Go-Moku.

PEOPLE: Two

ACTION:

Each person rolls a die. The highest roll goes first.

Player 1 rolls the dice and places them in the appropriate order (green first, then red) in the brackets below the grid. The green die determines the horizontal coordinate and the red die determines the vertical coordinate. The player then places her color token on the point which corresponds to those coordinates.

Play proceeds until a roll is repeated. When this happens the player may place her token on any free point.

The first person with 4 tokens in a row wins.

Coordinate Dice

(ditto sheet)

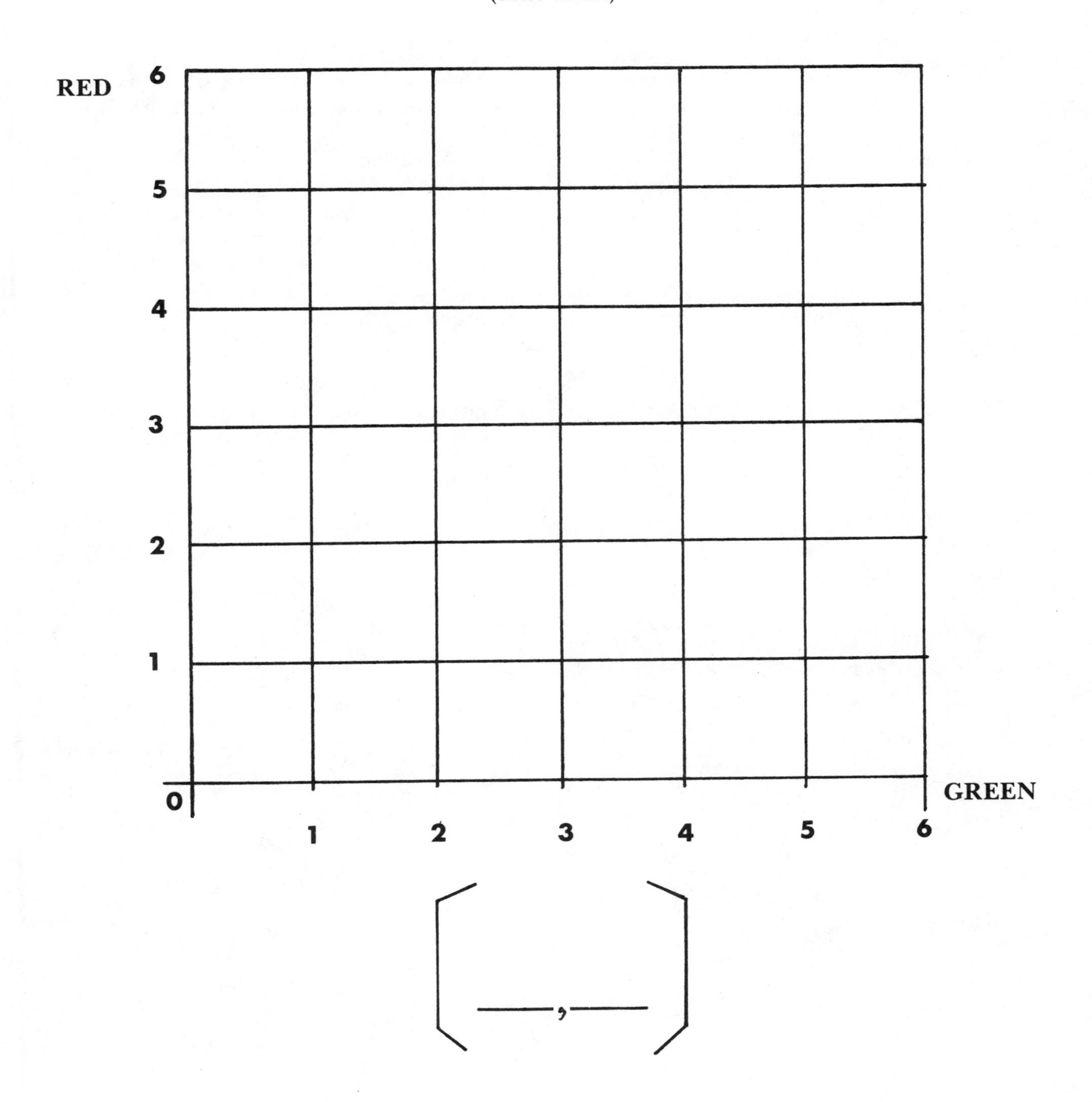

Which Job Do I Have?

- **Career awareness**

MATERIALS

- **List of jobs generated on the first day (see** *Jobs Men and Women Do***)**
- **3 x 5 cards**
- **Scratch paper**
- **Pencils**

Students become familiar with the different aspects and requirements of various careers by discussing them with their peers and attempting to articulate a description.

PEOPLE: 2-3

ACTION:

Have each student pick an occupation from the list generated on the first day (See *Jobs Men and Women Do*). Put a star by each one chosen. Each should write her choice on an index card. Collect all the index cards and shuffle them.

Divide the class into teams of 2-3. Each team draws a job title from the box.

Each team writes a job description for the job chosen. The description might include such characteristics as: works mainly indoors or outdoors; works with others or alone; travels a lot.

Each team then reads its job description and the rest of the group tries to guess the occupation it describes.

IF YOU LIKE:

Have the group generate a check list of job characteristics. Then create a composite description of each job by selecting descriptors from the list.

Day 6

Mystery Story

- **Group problem solving**
- **Breaking set**
- **Questioning strategies**

MATERIALS
- **None**

This activity stimulates creative thinking. Very often the student must break away from the solution path she has set for herself. Sometimes this can be accomplished by asking questions which lead in another direction entirely.

PEOPLE: Whole class

ACTION:

Tell the students you are going to ask them to solve a mystery. There is one rule — they may ask only questions which have yes or no answers. Any such question, however, is acceptable.

East or West

Tell the following story: Amy is sitting in her car. She starts it up and drives west in a straight line a quarter of a mile. When she stops the car, she is facing east. How can this be?

Solution: She backed the car up.

If the group runs out of questions, give hints about which facts are important.

IF YOU LIKE:

Have small groups make up mystery stories. Discuss what makes the mysteries difficult to solve. What are examples of other facts that might surprise us?

- unusually tall or short person.
- woman doing a "man's" job
- words with double meaning, such as, *a ring* is both an object and sound; *a bicycle* is both a vehicle and a deck of cards.

How Many Beans?

Students are challenged to invent ways of estimating the number of beans in a large jar without actually counting them. The situation presented is one in which the estimation is more appropriate than exact measurement. Strategies are introduced that aid estimation skills.

- **Estimation**
- **Sampling techniques**
- **Large numbers**

MATERIALS
- **Large jar of beans**
- **Containers of different sizes: liter, 25 m, 10 ml**
- **Graph paper of differing grid sizes**
- **Gram scales**
- **Masking tape**
- **Markers**
- **Butcher paper**
- **Pencils**
- **Paper**
- **Calculators (optional)**

PEOPLE: Whole class and small groups (2-3)

ACTION:

The challenge is to determine the number of beans in the large jar without counting them.

Each person guesses the number of beans she thinks are in the jar. These guesses are recorded and averaged by the instructor or one of the older students.

Present the following dilemma:

"You have been asked to order enough beans to feed 1,200 people at a banquet coming up next week. Each person will eat an average of 100 beans. How many of these jars of beans do you need to order? How can you make a decision without counting each bean?"

With the entire group, brainstorm possible sampling techniques. Here are some possibilities:

- Take a small container like a 25 ml cup and fill it with beans. Count them. Determine how many 250 ml cups of beans will fill the jar.
- Take some large-square graph paper, cover one square with beans and count them. Pour out the entire jar of beans onto 2 cm graph paper and see how many squares are covered.
- Use a gram scale to weigh 100 beans. Then weigh all the beans in the jar.

After the group has identified several methods of estimation, break the class into small groups to try 1 or 2 of the approaches described. Record all the estimates on the board and average them. Have the group examine the following questions: Which estimates are close to each other? Which methods seem most accurate? How close to the original guesses are the estimates? How accurate do you need to be when ordering beans?

Hurkle

- **Strategy**
- **Coordinates**
- **Compass directions**

MATERIALS
- **Hurkle paper**
- **Pencils**
- **Blackboard and chalk for demonstrations**

The search strategy in "Hurkle" is a two dimensional extension of the binary search strategy used in "Guess". Coordinate and compass directions are used in the hunt for the hurkle.

PEOPLE: Small group

ACTION:

Pass out hurkle paper. (Familiarize students with the coordinate names of the points and determine whether or not they know compass directions like North, South, East, West, Northwest, etc.)

Play one or two practice games at the blackboard with the whole class before dividing students into small groups.

Explain to the group that a hurkle is a creature hiding on a 10 x 10 grid. You can only find it by guessing the point where it is hiding. Each point has a coordinate name consisting of 2 numbers. The first number tells which column (how far across), and the second number tells which row (how far up). Each turn, players try to guess the coordinate name of the point on which the hurkle is sitting. After each guess, the leader tells the group the compass direction of the hurkle's hiding place relative to the guess. Each guess should be written down. For example: If the hurkle is hiding at (5,6) and the guess is (1,2), the clue is "The hurkle is hiding northeast of there."

IF YOU LIKE:

After the students have found the hurkle several times, see if they can decrease the number of guesses needed to find the hurkle. The strategy often surfaces as students attempt to do this.

Hurkle Paper

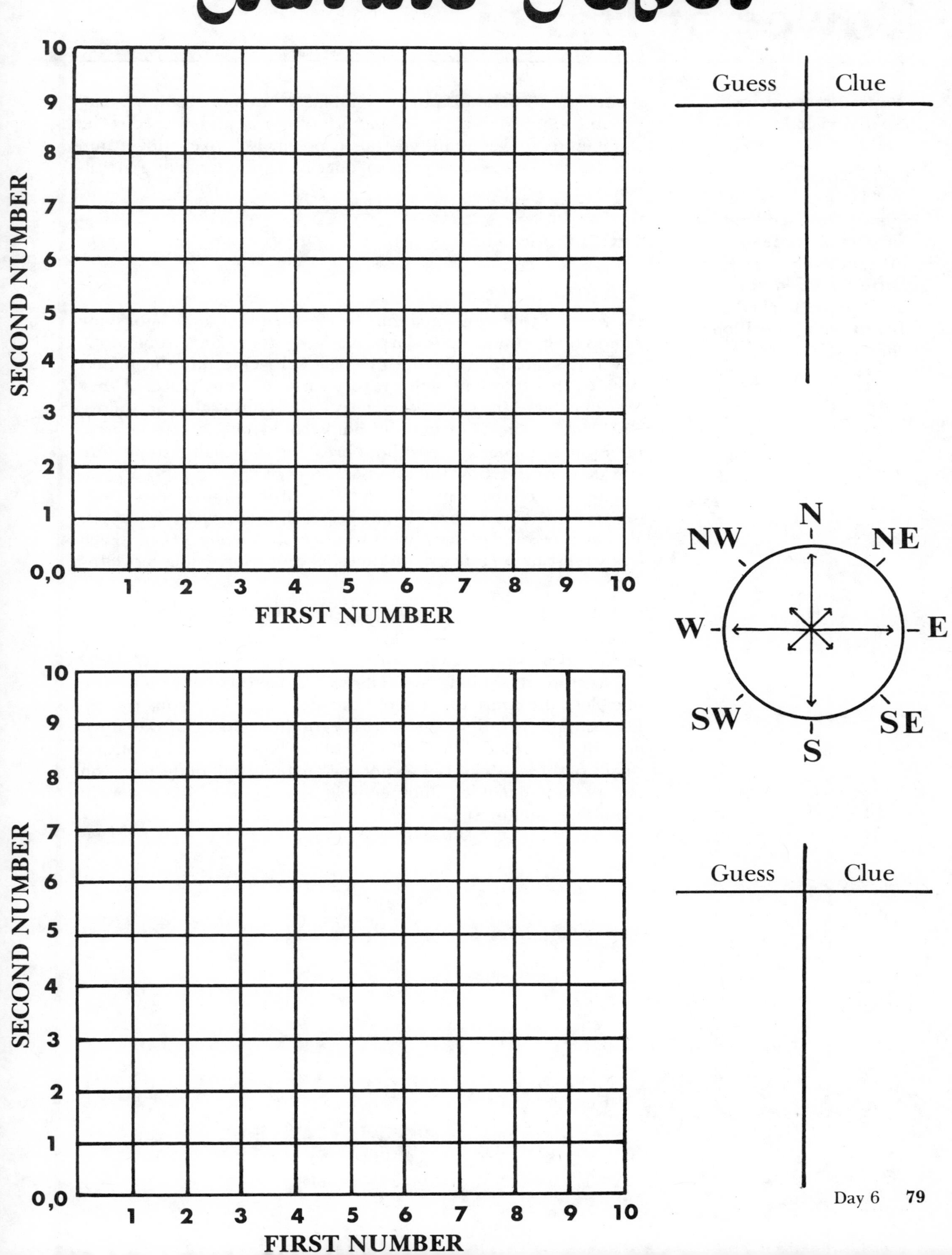

Tower of Hanoi

- **Pattern recognition**
- **Spatial visualization**

MATERIALS
- **A "Tower of Hanoi" puzzle for each student or pair of students. Alternatively, use 3 different sized blocks that can be stacked and a board with 3 positions marked.**

The "Tower of Hanoi" puzzle is similar to but more complex than "Leap Frog". To uncover the pattern of moves and to predict the number of moves required, the problem is broken down into simpler versions and the pattern, once recognized, can be generalized.

PEOPLE: Individuals or pairs

ACTION:

Present the following situation: In Hanoi, a long time ago, some monks built a tower out of large blocks of stone. Each stone of the tower was smaller than the one below it. One day the monks decided that they had built the tower in the wrong place, so they moved it. Since the stones were heavy, the monks could only move one block at a time. Further, the monks could not put a larger stone on top of a smaller stone without breaking the smaller stone. Can you demonstrate how the tower was moved?

The wooden tower provides a model of the monks' tower. Students must move the tower from one peg to another, *moving only one block at a time* and *always placing smaller blocks on larger blocks.* Each block must be on a peg at all times, except while being moved.

STRATEGY:

After the students have struggled for awhile, suggest that they build and move a smaller tower (e.g., one with only two blocks).

The students should eventually realize that each time they move one block the entire tower above the block must be reconstructed. For example, in order to move a tower of three blocks, a tower must be built of one block (the top block), of two blocks (the top two), and finally of three blocks. It does not work to build half of the tower on one pole and half on another pole as an intermediate step in rebuilding the tower.

IF YOU LIKE:

After the students understand how to move the tower from one pole to another, have them discover how many moves it takes. Again, the strategy is to make a simpler problem.

Have students begin with one and two block towers. They should gradually increase the number of blocks until they see a pattern. The pattern should lead students to the number of moves it takes to move the whole tower.

Making a chart of the number of blocks in a tower and the number of moves required may facilitate this discovery. Keep a column for the students to predict the number for each tower.

# of Blocks	# of Moves	Guess # of Moves
1	1	?
2	3	?
3	7	?
4	—	?

Encourage the students to come up with a rule or function that would generate the number of the moves for each tower size. The function is (2^n-1), where n is the number of blocks. Another description of the function is: Twice the last number of moves plus one.

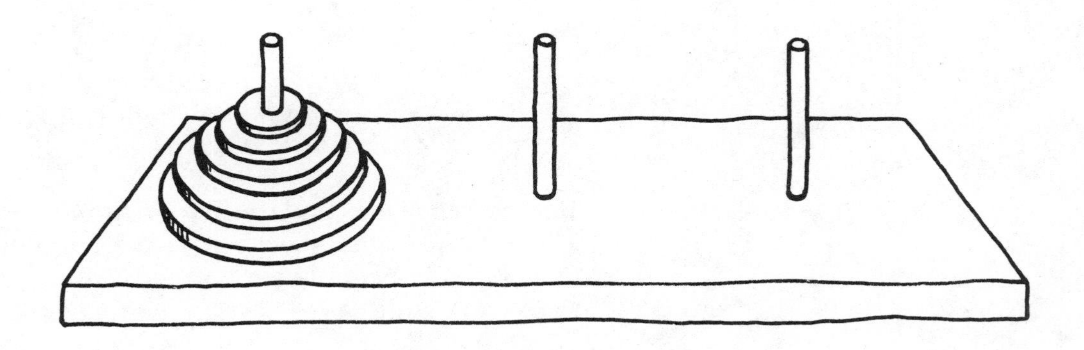

Fiddle Faddle Flop

- **Logical reasoning**
- **Strategy**
- **2 and 3 digit numbers**

MATERIALS
- **Blackboard**
- **Chalk**

The hidden number is found by a process of elimination. Each guess gives new clues to the hidden number. These are recorded in a table. As students become skillful at guessing, the number of guesses needed to determine the hidden number decreases.

PEOPLE: Whole class

ACTION:

A leader thinks of a secret two-digit number in which the digits are not the same. The group attempts to determine the number by guessing two-digit numbers.

After each guess, the leader responds:

Fiddle—If one digit is correct and also in the correct place.
Faddle—If one digit matches but is in the wrong place.
Flop—If no digits in the guessed number match those in the secret number.
Faddle Faddle, for example, would indicate that two digits are correct and both are in the wrong place.

Here's a sample game:

Guess	Leader's Response
63	Flop
25	Faddle
94	Flop
27	Fiddle
75	Faddle Faddle
57	Fiddle Fiddle. That's it!

The challenge is to find the hidden number in the fewest possible guesses. To do this, players need to guess numbers which will eliminate many possibilities.

Encourage students to devise ways to keep track of the numbers eliminated with each clue.

IF YOU LIKE:

- Use three-digit numbers.
- Have the group play *Mastermind.*
- Use three-letter words in place of numbers Do not repeat any letter.

This Is Your Life

It is easy to step out into the working world with many misconceptions. Most students are unaware that 9 out of 10 women work and women work an average of 22 years during their lifetime. These statistics are personalized as the students attempt to form a life timeline.

PEOPLE: Individuals

ACTION:

Give each student eight squares of paper. They label the first one 0-10 years, the second 10-20 years, and continue to 70-80 years.

On each square of paper the student attempts to describe or predict what her life is or will be like during those years. This may be done with words or pictures or both. The following questions are samples of those the students might think about while creating their timeline:

- How old are people when they start a family?
- Will you work? For how long?
- What exciting things will you do?
- When will you get your first car?

Once everyone has finished their timeline, have them each share a significant event in each decade with the rest of the group. Then hang the timelines on the wall.

- **Career awareness**
- **Work life consciousness— age, money**

MATERIALS

- **Squares of paper**
- **Pencils**
- **Crayons**

Build a Structure

- **Spatial visualization in 3 dimensions**
- **Intuitive geometry and physics**
- **Creative thinking**
- **Cooperation**

MATERIALS

- **Construct-O straws or D-stix set for each group. Alternately, plastic straws, string, scissors and tape can be used.**

Students have the opportunity to create any structure they want. With the goal of building a tall tower, students can use exploration and intuition to determine what shapes are strong and rigid.

PEOPLE: Groups of 3-4

ACTION:

Ask students to close their eyes and visualize a tall tower, one that holds a bridge or a heavy telephone wire. What shape is the tower? How tall is it? What makes it strong? Can you see the bracing?

Then have students open their eyes and try to build a tall tower, using the materials listed. Each group should build the tallest tower they can.

When they have finished, the tower must be able to stand alone for at least 1 minute.

At the end of the activity, have each group talk about their success or failure. Mention career links to the modelling of architects and engineers

IF YOU LIKE:

- Build a bridge that will span the widest gap.
- Use plastic straws and pins or tape. The challenge for each group is to build the tallest structure they can. Once each group has completed a structure, they each start to cut straws in half. See which group can make the most cuts before their structure collapses.

Day 7

Mystery Stories

- **Group problem solving**
- **Breaking set**
- **Questioning strategies**

MATERIALS
- **None**

This activity stimulates creative thinking. Very often the student must break away from the solution path she has set for herself. Sometimes this can be accomplished by asking questions which lead in another direction entirely.

PEOPLE: Whole class

ACTION:

Tell the students you are going to ask them to solve a mystery. There is one rule — they may ask only questions which have yes or no answers. Any such question, however, is acceptable.

Walk on Water

Tell the following story: Marcia was a Magician by trade. However, her fans were hard pressed to believe she would be able to perform the advertised feat — walking on water. The lake on the edge of town was chosen for the event. At the announced time, Marcia did walk on the surface of the water. How did she do it?

Solution: Marcia's fans walked on the water along with her. In fact, they had an ice-skating party on the frozen lake.

Enlightening Tale

Tell the following story: Elsa was very clever. One of her tricks was to turn off the light in her bedroom and get into her bed which was five feet away from the light switch, before the room was dark. How did she do it?

Solution: It was still daylight when she went to bed.

IF YOU LIKE:

Have small groups make up mystery stories. Discuss what makes the mysteries difficult to solve? What are examples of other facts that might surprise us?

- unusually tall or short person
- woman doing a "man's" job
- words with double meaning, such as, *a ring* is both an object and sound; *a bicycle* is both a vehicle and a deck of cards.

Guess the Carob Balls

This is a sequel to "How Many Frogs?" and "How Many Beans?" Again, the students are encouraged to guess. Sampling techniques can be applied to improve estimates. Guesses can be graphed and the range, frequency, median, mean, and error can be discussed.

- **Guessing**
- **Estimating**
- **Recording data**
- **Concepts of range, frequency, median, and mode**

PEOPLE: Two

MATERIALS
- **Jar of between 100 and 1000 carob balls**
- **Ways to record guesses: 100-square numbered 10 to 1000 by tens, covered with clear contact**
- **Counters such as pennies or squares of cardboard**

ACTION:

Have the class work in pairs. This activity is similar to "How Many Frogs?" except that the number of carob balls is over 100 while the number of frogs was under 100. See "How Many Frogs?" for details of implementing the guessing and recording of guesses. The students first make guesses, then they can apply techniques learned in previous activities to revise their estimates. Count the carob balls. Compare original guesses with revised estimates.

10	20	30	40	50	60	70	80	90	100
110	120	130	140	150	160	170	180	190	200
210	220	230	240	250	260	270	280	290	300
310	320	330	340	350	360	370	380	390	400
410	420	430	440	450	460	470	480	490	500
510	520	530	540	550	560	570	580	590	600
610	620	630	640	650	660	670	680	690	700

IF YOU LIKE:

Estimation activities can also focus on measures of length (how many meters long is the floor?), or of time (guess when a minute is up).

Span

- **Strategy**
- **Data collection**

MATERIALS
- **9 pennies**
- **A 3 x 3 grid**

This game is fast and always has a winner. Therefore, students will play many times. In this way students can collect much data from which to deduce the optimum strategy. In the case of beginning players, the first person to play usually wins more often than the second player. As the strategy emerges, this situation changes.

PEOPLE: Two

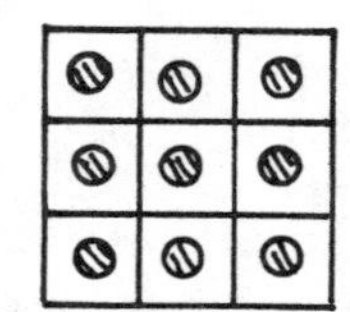

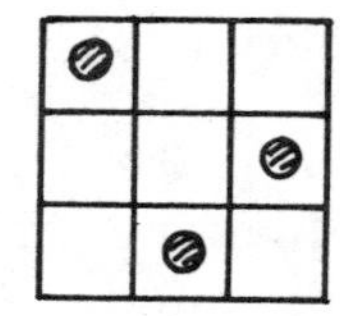

ACTION:

Players place one penny in each square of a 3 x 3 grid. Each player at her turn removes one penny from any square she chooses. She must leave at least one penny in each row and each column. The last person to play wins. If a player removes a penny so that one row or column is blank, she loses.

Let the loser of each game decide whether to play first or second in the next game. One way for students to start to crack the strategy is to play a series of at least 10 games. At the end of each game, they record:

1. Who won — the first player or the second?
2. The number of coins remaining at the end.

1. Who won?	2. How many pennies left?

Players might also make note of the patterns made by the pennies that are left on the board at the end of the game.

IF YOU LIKE:

- Each player may remove as many adjacent pennies from one row or column as she wishes on one turn.
- Start with 16 pennies on a 4 x 4 grid.

Inventions

Students are given an outlandish contraption to invent. It should be outlandish enough so that none has a preconceived idea of what the end product looks like. They can describe their inventions visually (through pictures) or verbally (through a written description).

- **Creative thinking**
- **Art**
- **General concepts of science, math**

MATERIALS
- **Paper**
- **Pencils**
- **Drawing paper**
- **Crayons or felt pens**

PEOPLE: Individuals

ACTION:

The teacher presents a list of possible inventions from which each person selects one that interests her. Some possibilites are:

- A fun machine.
- A way to weigh an elephant.
- A weather machine.
- A way to count forever.
- A perfect robot.

Have each student draw a picture of her own invention.

Have students share their inventions and pictures with the rest of the class.

IF YOU LIKE:

Have each person describe her invention to a partner and have that partner draw a picture of the invention.

Where's the Rectangle?

- **Strategy**
- **Coordinates**
- **Area**

MATERIALS

- **8 x 8 Coordinate grid for each player (can be re-used if covered with plastic)**
- **Pencils**

Once students become familiar with coordinates, they can use this knowlege in strategy games such as "Where's the Rectangle?" The strategy here depends on recognizing the different rectangles that can be drawn with a fixed area.

PEOPLE: 3-5

ACTION:

Remind students of the coordinate names for points. Review concept of area as "squares inside" a rectangle.

Divide into groups of 3-5 students. Each person needs a blank grid card and a pencil.

One person draws a rectangle on her grid without showing the other players. Its sides must coincide with grid lines. See the example below. She then tells the rest of the group the area of her rectangle.

Other players try to locate the four corners of the hidden rectangle. Each player in turn guesses a coordinate point. The person hiding the rectangle tells whether that point is on, in, or out of her rectangle. These clues are recorded.

Play continues until the corner points are identified. The number of questions needed is the group score.

Let someone else hide a rectangle. This time the group tries to reduce the number of guesses required to identify the four corners.

Initial Clue: The area is 16.

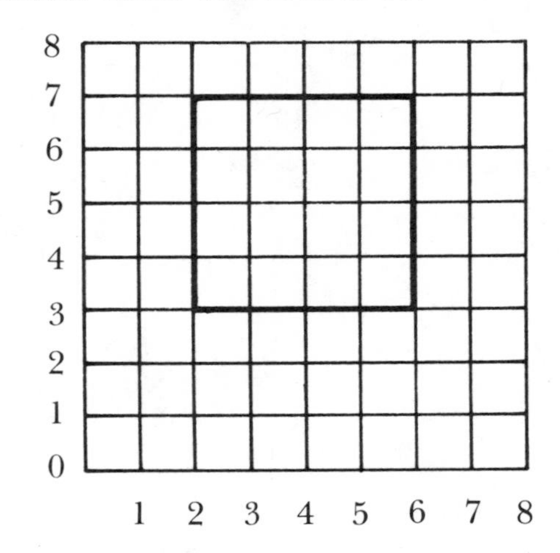

Guess	Response
(5,6)	in
(7,2)	out
(6,4))	on

IF YOU LIKE:

- Fix an area (say 24) and have students find all the different shaped rectangles with that area. Which areas could describe different rectangles and which describe only a few?
- Let students decide if they would rather know the area, the perimeter, or the length of one side as an initial clue.

Where's the Rectangle?

(ditto sheet)

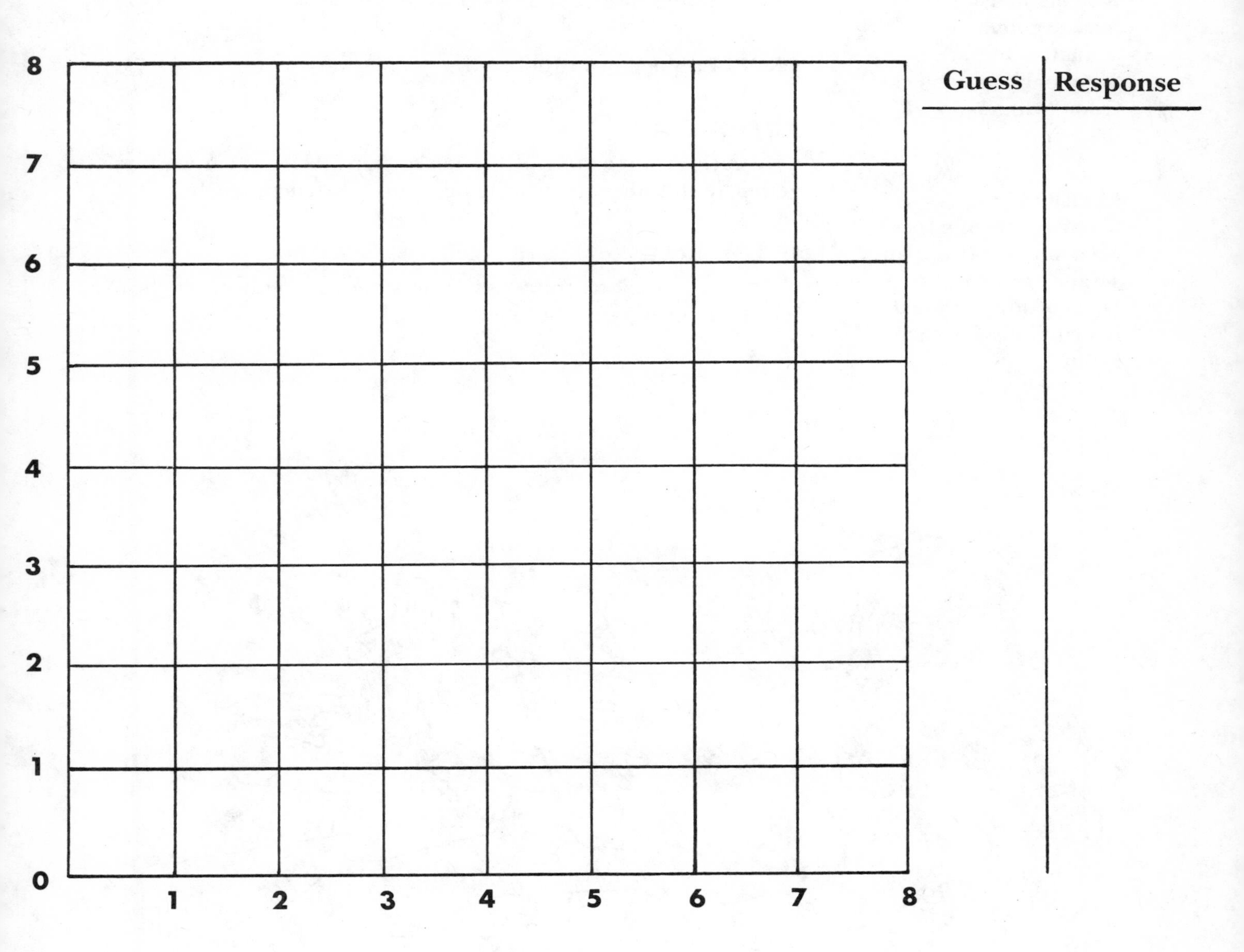

Origami Box

- **Symmetry**
- **Non-standard measurement**
- **Visualization of 3-dimensional objects from 2-dimensional diagrams**

MATERIALS

- **Square sheet of paper for each student (wallpaper scraps are nice!)**
- **Ditto of diagrammed instructions for each group**

In this activity each student constructs a box using only symmetric paper-folding to design and measure her box.

PEOPLE: Small groups (2 or 3)

ACTION:

Give each group a copy of the instructions. Have them work together to decipher the instructions and construct their boxes.

IF YOU LIKE:

Have one leader attempt to give the paper folding instructions to the whole group or to a small group.

ORIGAMI BOX

(ditto sheet)

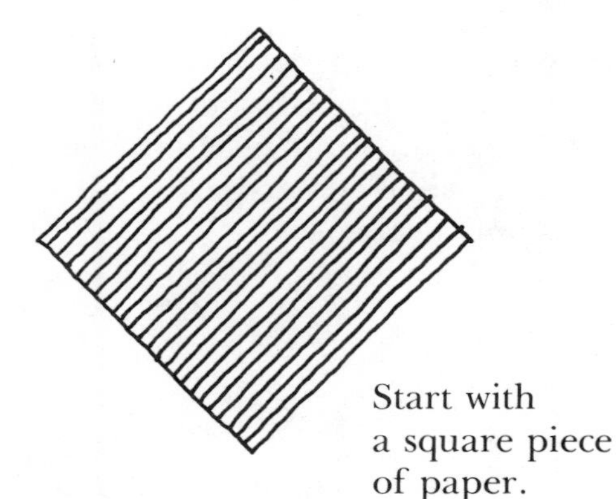

Start with a square piece of paper.

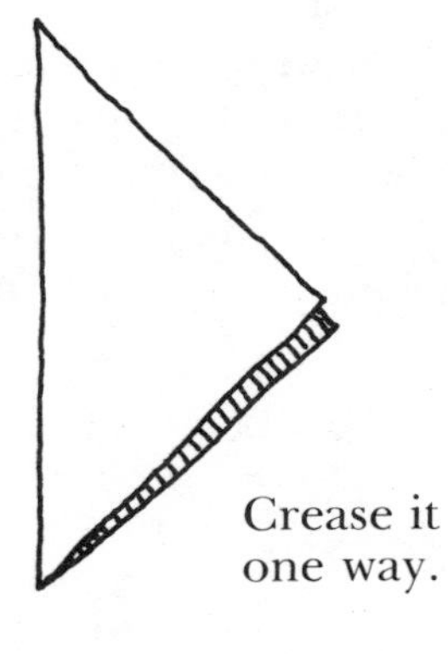

Crease it one way.

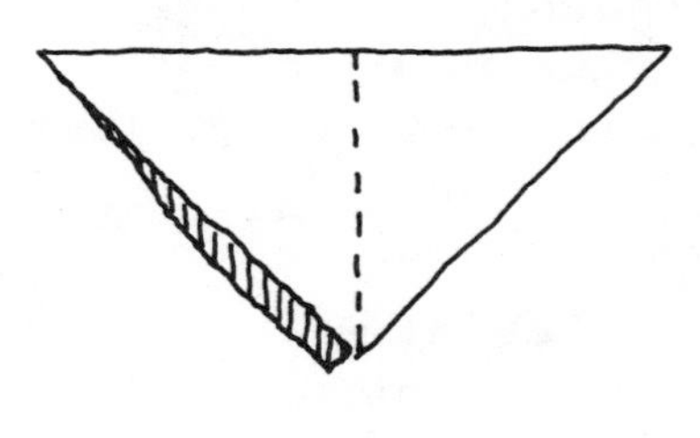

Crease it the other way.

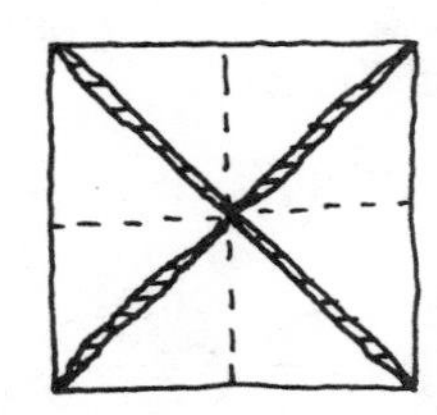

Fold all four points in to center.

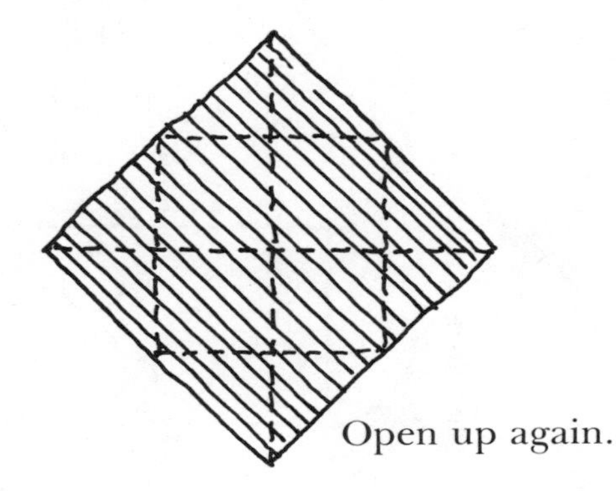

Open up again.

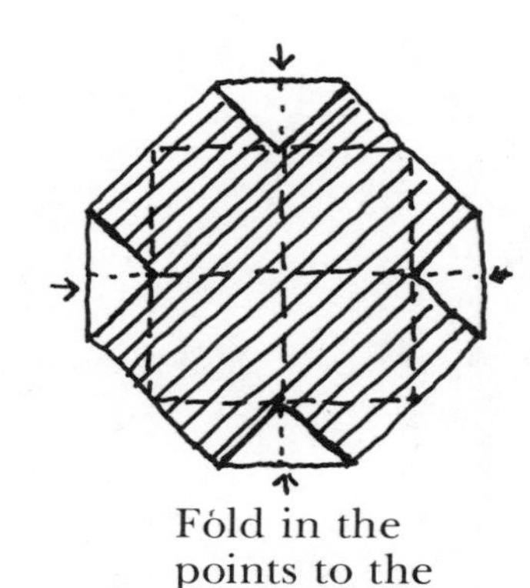

Fold in the points to the first fold line.

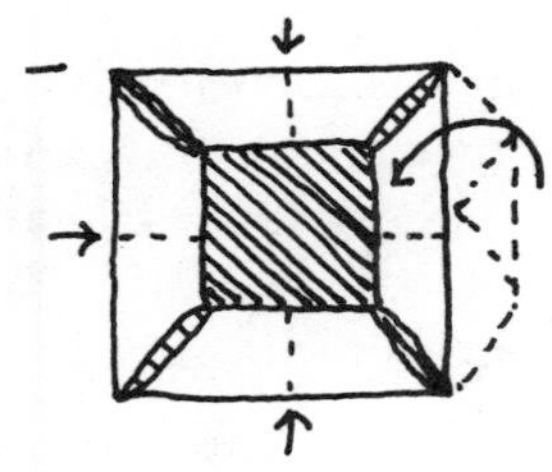

Fold in the sides where you just folded in the points — fold on the fold lines already there.

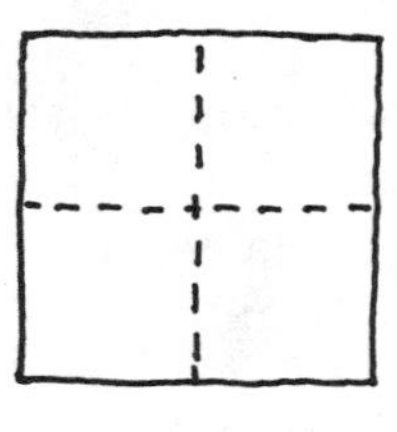

Turn it over.

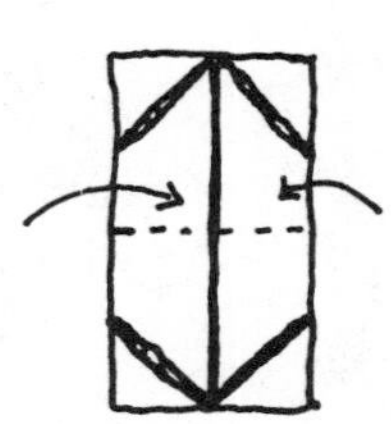

Fold right and left sides up so they meet in the center on top.

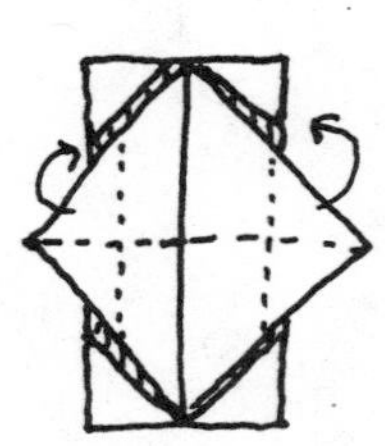

Lift those flaps up in the air.

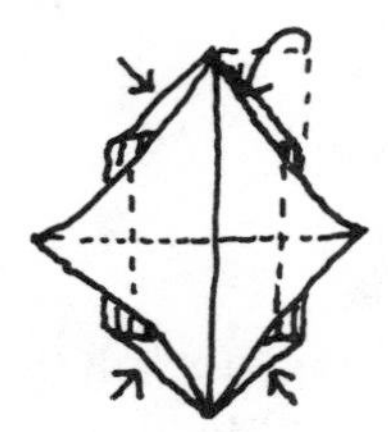

Fold all four corners up and in.

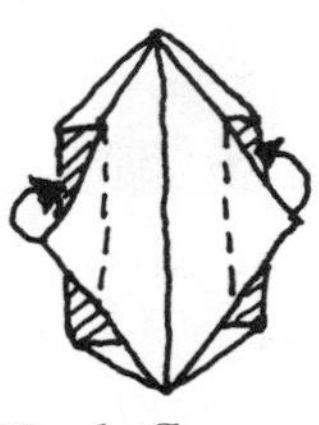

Tuck flaps to inside and flatten entire figure.

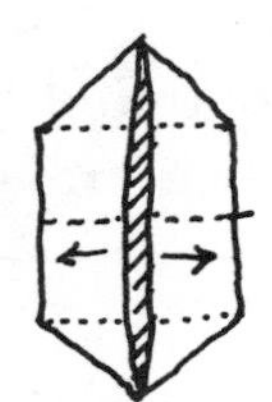

Open the center slot. Pull the two sides apart to form a box, sharpening folds as needed.

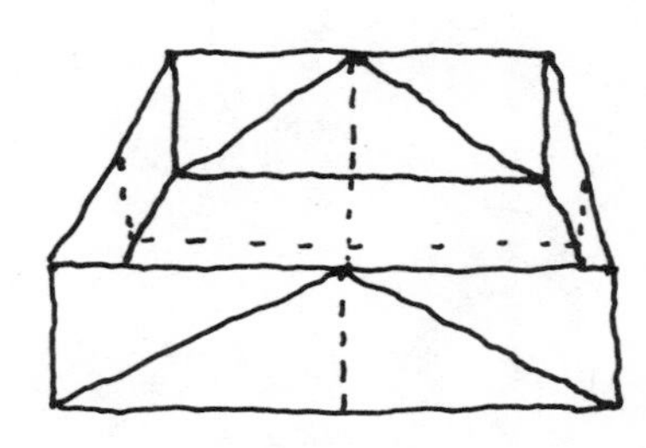

Kalah

- **Strategy**
- **Planning ahead**
- **Predicting moves of the opponent**

MATERIALS
- **Kalah board***
- **Beans (or other counters)**

This strategy game is a little different from others in this book. Many strategy games, like Guess, have one simple optimum strategy. "Kalah", however, requires a change in strategy after every move. This gives students experience in altering perspective on a problem and in re-examining plans and strategies.

PEOPLE: Two

ACTION:
Each player sits on one side of the Kalah board. The six pits on her side belong to her as does the Kalah to her right. Put three beans in each pit and none in either Kalah to start the game. Players try to collect as many beans as possible in their Kalah.

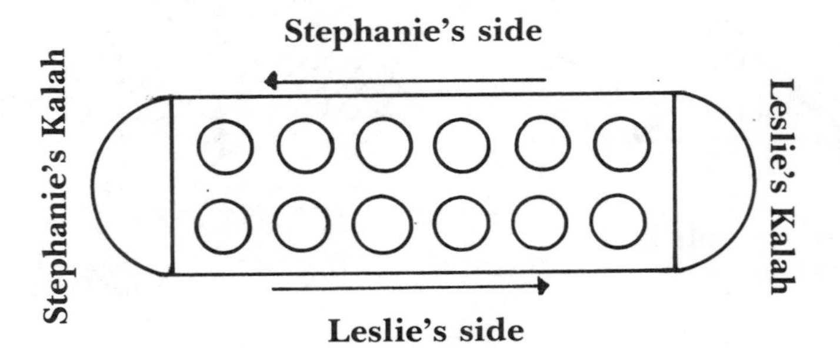

Each person in turn picks up all the beans in any one of her own six pits and places them one by one in each pit around the board to the right, including, if there are enough, her own Kalah and on into her opponent's pits (but not her opponent's Kalah). Once a bean is in a Kalah, it stays there.

If her last bean lands in her own Kalah, she gets another turn.

If her last bean lands in an empty pit *on her side,* she captures all her opponent's beans in the opposite pit and puts them in her Kalah together with the capturing bean. The game ends when all six pits on one side are empty.

IF YOU LIKE:
- After students have played several games, discuss strategies they have found helpful.
- Try different numbers of beans in the pits.
- Make a large chart of the directions above.

*A Kalah board can be made out of an egg carton. Cut off the lid, cut the lid in half and fasten each half to one end of the carton. This will form the two Kalahs.

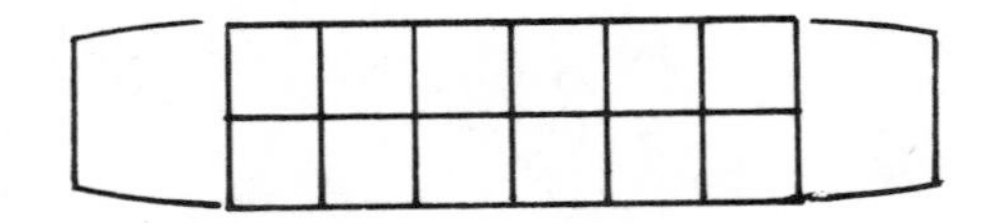

Day 8

Symmetric Art

- **Symmetry**
- **Proportionality**
- **Art**

MATERIALS
- **Magazine pictures of half-faces glued on drawing paper**
- **Paper**
- **Crayons**
- **Pencils**

Symmetry and proportionality are two concepts important both in mathematics and in art. In the first activity, students work with their awareness of symmetry in the human body by attempting to draw the missing half of a face. In the second activity, students each draw one quadrant of a picture of a face. As the students attempt to put the four pieces together, the importance of proportionality becomes evident.

PEOPLE: Individuals and Groups of 4

ACTION:

Finish the Face

Before class, collect life-sized or reasonably large pictures of faces from magazines. Cut them in half vertically. Glue one side to a piece of 8½ x 11 white drawing paper. let each student select a face and draw in the other half.

MATERIALS
- **Large pictures of faces or objects glued to a backing and marked off in quadrants**
- **Blank drawing paper**
- **Pencils**

Quadrants

Divide the class into four groups of four. Give each group one of the pictures. Assign each individual in the group to draw one of the quadrants of the picture. Do not give them further instructions and do not let them know in advance what will be done with the finished quadrants.

After everyone has drawn her quadrant, have the group tape the pieces together to form the complete face.

Discuss the results. How could the quadrants have matched better?

Double Digit

Both skill and chance play important roles in this game. The dice rolls make it difficult to implement a consistent winning strategy. However, an intuitive understanding of probability will allow students to find a strategy that will be successful more often than not. Estimation skills will aid a student's chances for success.

- **Strategy**
- **Estimation**
- **Place value**
- **Probability**

MATERIALS
- **A die**
- **Scoresheets**

PEOPLE: 2-6

ACTION:
Each group has a die and a scoresheet for each person like the one below.

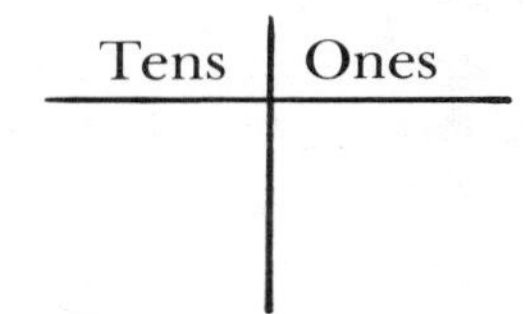

The members of the group take turns rolling the die. Each player places the number rolled in either the tens column or the ones column of her score sheet. When a number is entered in the tens column, players put a "0" next to it in the ones. Thus, a 4 in the tens column counts as 40.

After the die has been rolled 7 times, each player adds up her numbers. If a player's total is over 100, she is out of the game.

The players left in the game compare their totals. The player who's closest to 100 gets a point. If players are tied, they each get a point.

Play 5 games. The player with the highest score at the end of 5 games wins.

IF YOU LIKE:
- At the end of the game, ask the students to figure out what the best total could have been with those seven rolls.
- **Harder version:** Players roll the die seven times. The total score may not exceed 60 and the closest score to 60 wins.

Multi-Color

- **Strategy**
- **Area**
- **Coordinates**

MATERIALS

- **10 x 10 grid with spaces numbered along the bottom and left side. (Note the difference from numbered lines)**
- **Pencils**
- **Crayons (optional)**

This is an extension of "Find the Rectangle." Students try to determine the location of the 4 colored rectangles filling a 10 x 10 grid. The object is to find them with the minimum number of questions.

PEOPLE: 2-6

ACTION:

Each player needs a numbered grid and a pencil. The leader divides her grid into four rectangles any way she wants. However, no rectangle may have a side of length "one." Without revealing the shapes of the rectangles she colors or labels each one a different color.

Finally, she tells the other players the area of the four rectangles.

In "Multi-Color" each *square* on the grid has a coordinate name. Players take turns giving coordinates to ask about the color of a square. The first number of the pair refers to the number of the column containing the square; the second number refers to the number of the row containing the square. For example: (8,3) refers to the little square with the "x" on it, on the sample grid below. When a player thinks she has replicated the leader's grid, she shows it to the leader. The game continues until someone successfully duplicates the leader's grid.
Here is an example:

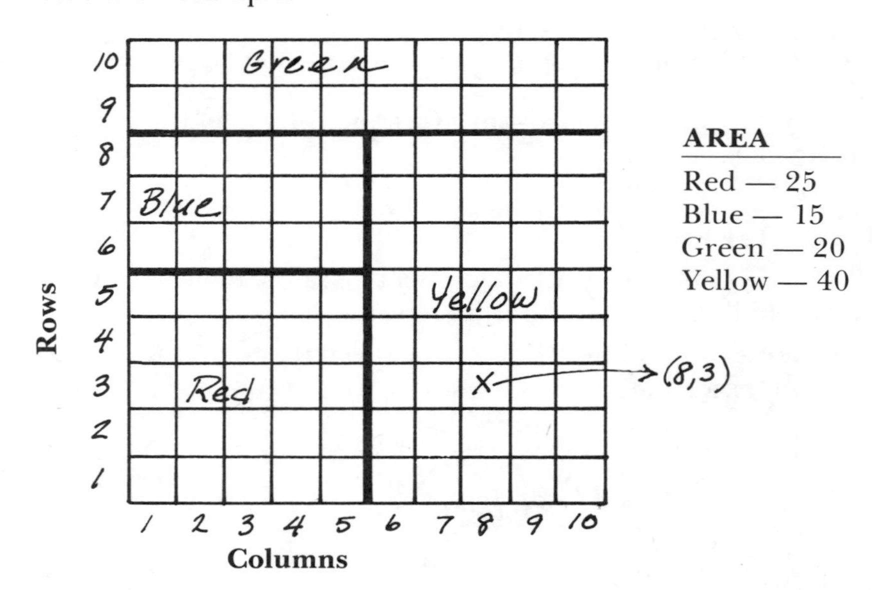

IF YOU LIKE:

Have students work in pairs. Each person can color her own 10 x 10 grid. The two players then take turns asking about the other's grid. The first one to correctly guess the other's pattern and color of rectangles wins.

Sticks

- **Strategy**
- **Problem solving**

MATERIALS
- **Pencils**
- **Scratch paper**

This game is similar in play and strategy to the "Nim" games that appeared earlier. Finding the winning strategy for "Sticks" involves students in problem solving, techniques like thinking ahead, working backwards, and simplifying the problem.

PEOPLE: Two

ACTION:
Draw 7 sticks in a row.

Each player in turn crosses off 2 adjacent sticks.

The last person who can play wins. There will often be some single isolated sticks left at the end of the game.

First play won this time.

IF YOU LIKE:
Vary the number of sticks layed out at the beginning of the game. Play each version enough times to determine if the first player or the second player should win. Record your data on a chart like this:

Sticks at start	1	2	3	4	5	6	7	8	9	10	11	12
Winner (First or Second Player)												

Build a Puzzle

- **Spatial visualization**
- **Geometry**
- **Art**
- **Symmetry**

MATERIALS
- **Puzzle patterns**
- **Scissors**
- **Tape**
- **A model of a tetrahedron**

In this activity, students create and put together their own three-dimensional puzzle. Each three-dimensional puzzle piece is constructed from a two-dimensional pattern. In attempting to fit the pieces together to form a regular tetrahedron, students need to visualize the effects of rotating and reflecting the pieces.

PEOPLE: Individuals working in small groups

ACTION:
Students each cut out two pattern pieces. By folding each piece on all lines and taping it together, two identical three-dimensional solids can be formed.

Using the model tetrahedron as a visual guide, students try to place the two solids together in such a way that a regular tetrahedron is formed.

IF YOU LIKE:
Have students put together a three-dimensional puzzle that forms a cube. Patterns for these shapes are included.

To make a model of a tetrahedron
Cut out around the outside edge. Fold on the inside lines and tape edges together to make a solid shape.

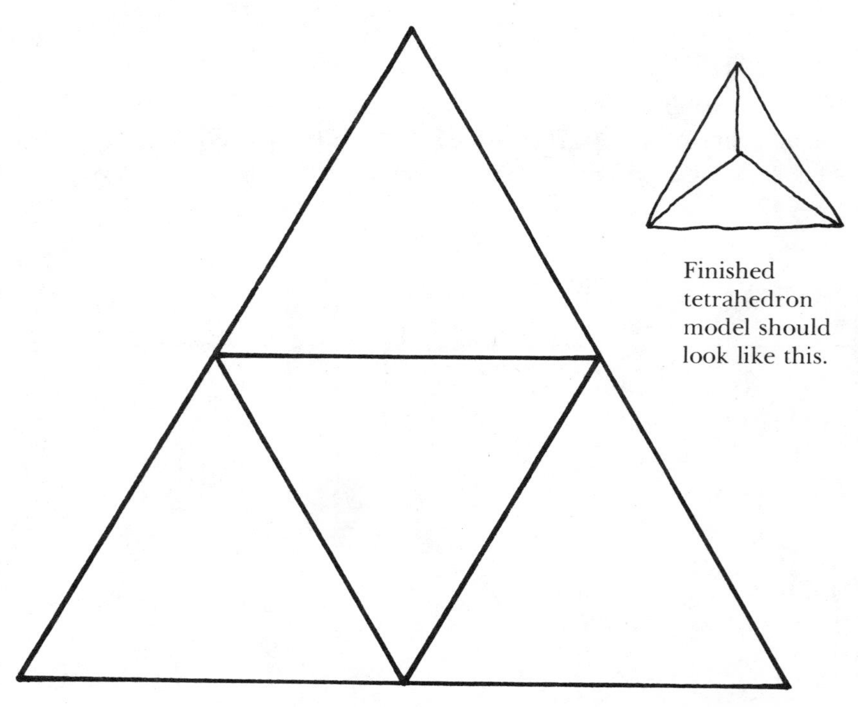

Tetrahedron Pattern (1 piece)

PUZZLE PIECES

(ditto sheet)

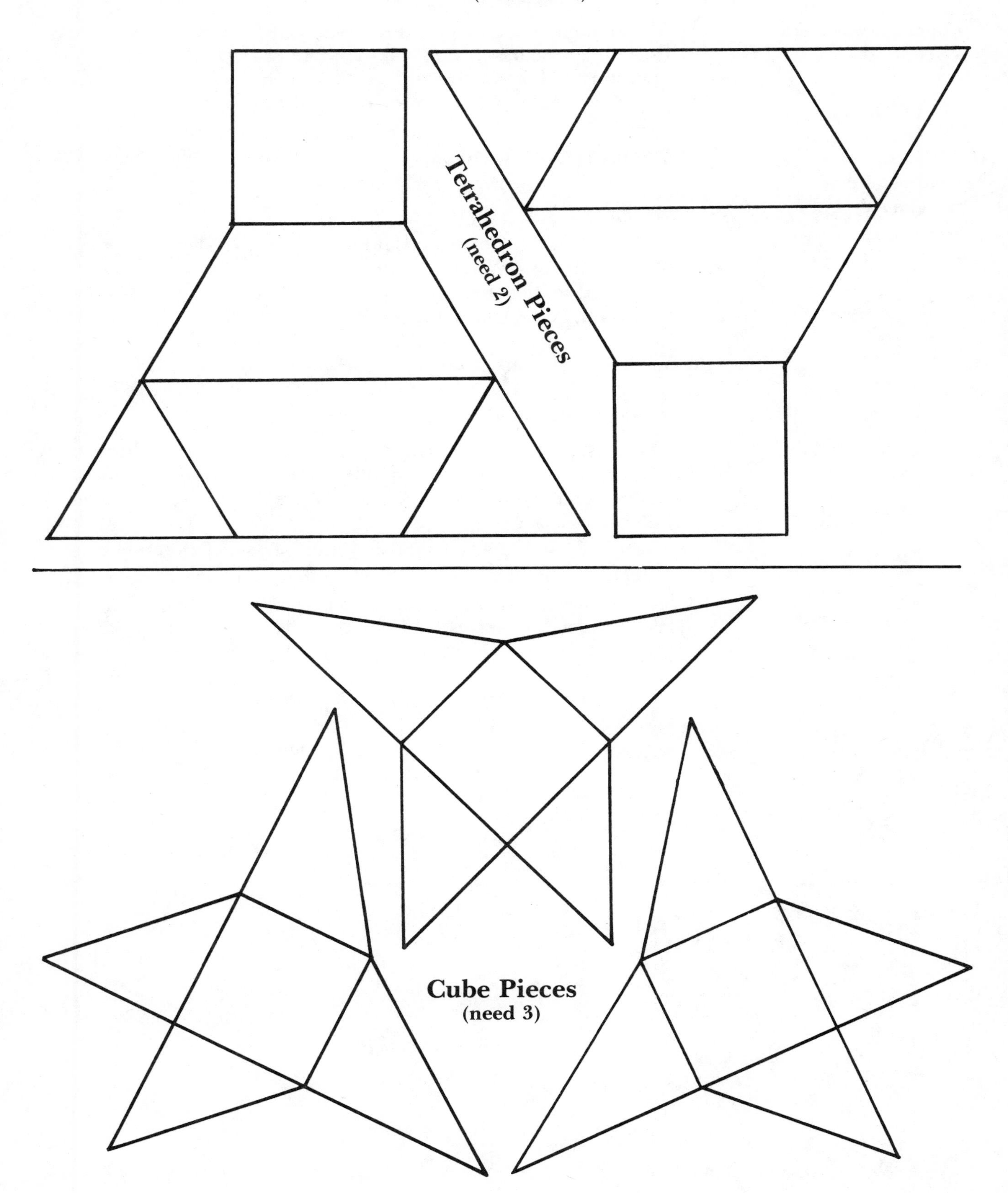

Collage Of Women Working

- **Career awareness**
- **Art**

MATERIALS
- **Pictures**
- **Scissors**
- **Rubber Cement**
- **Tagboard**

The media are increasingly filled with articles and advertisements depicting women working in a wide range of career situations. By collecting and displaying pictures of women working, students can broaden their awareness of career options.

PEOPLE: Whole class

ACTION:

Throughout the course, encourage students to collect and bring to class pictures of women working.

Challenge them to bring in pictures of women working in occupations that are unusual for women.

The last day of class, let students design a collage with their picture collection.

IF YOU LIKE:

- Let each student pick a favorite picture and explore the illustrated occupation further. She can look up the educational requirements, employment outlook, and salary in the *Occupational Outlook Handbook.*
- Arrange for students to take shapshots of women they encounter engaged in non-traditional work.
- Have students collect pictures of men working in occupations that are unusual for males, and make a collage with the collection.

Mystery Stories

- **Group problem solving**
- **Breaking set**
- **Questioning strategies**

MATERIALS
- **None**

This activity stimulates creative thinking. Very often the student must break away from the solution path she has set for herself. Sometimes this can be accomplished by asking questions which lead in another direction entirely.

PEOPLE: Whole class

ACTION:

Tell the students you are going to ask them to solve a mystery. There is one rule — they may ask only questions which have yes or no answers. Any such question, however, is acceptable.

Train Tunnel

Tell the following story: Two train tracks run parallel except for a spot where they go through a tunnel. The tunnel is not wide enough to accommodate both tracks, so they become a single track for the distance of the tunnel.

One afternoon a train entered the tunnel going in one direction, and another train entered the same tunnel going in the opposite direction. How did they avoid the accident?

Solution: They went through the same tunnel on the same afternoon, but one went through at 2:00 p.m. and the other went through at 3:00 p.m.

Winning Team

Tell the following story: Last week I went to a baseball game and my favorite team, the Lions, beat their opponents, the Tigers, 14 to 3. They managed to do this without a single man on the Lions' team scoring a run. Can you figure out how?

Solution: The Lions' team was a women's team.

Sugar in the Coffee

Tell the following story: One morning Harry dropped a sugar cube in his coffee. A minute later he lifted the sugar cube out intact. How did he manage to do this?

Solution: Yes, it was a real sugar cube and real coffee. However, the coffee had not been made yet. It was instant coffee.

IF YOU LIKE:

Have small groups make up mystery stories. Discuss what makes the mysteries difficult to solve. What are examples of other facts that might surprise us?

Alphabetical Listing of Activities

Bibliography and List of Resources

Mystery Stories and Other Set Breaking Activities:

Brooke, Maxey. *Coin Games and Puzzles*. 1973. *Tricks, Games, and Puzzles with Matches*. 1973. Dover Publications, Inc. 180 Varick St., New York, NY 10014.

Burns, Marilyn, *The Book of Think*. 1976. *The I Hate Mathematics Book*. 1975. Little, Brown & Co., 1200 West St., Waltham, Mass. 02154.

Gardiner, Martin. *Aha! Insight*. 1978. Scientific American, W.H. Freeman & Co., 660 Market St., San Francisco, CA 94104.

Classification, Logical Thinking, Patterns and Functions:

Anderson, Carolyn and Jackie Haller. *Brain Stretchers,* Books I and II. 1975. Midwest Publications, P.O. Box 129, Troy, Michigan 48099.

Baratta-Lorton, Mary. *Mathematics Their Way*. 1976. Addison-Wesley Publishing Co., Sand Hill Rd., Menlo Park, CA 94025.

Burns, Marilyn. See above.

Burns, Marilyn. *Good Time Math Event Book*. 1977. Creative Publications, P.O. Box 10328, Palo Alto, CA 94303.

Elementary Science Study Teacher Guides. 1967-1970. Webster Division, McGraw-Hill Book Company, 1221 Avenue of the Americas, New York, NY 10020. See especially: *Attribute Games and Problems* and *Pattern Blocks*.

Greenes, Carole, *Problem-Mathics*. 1977. Creative Publications, P.O. Box 10328, Palo Alto, CA 94303.

Greenes, Carole, John Gregory, and Dale Seymour. *Successful Problem Solving Techniques*. 1977. P.O. Box 10328, Palo Alto, CA 94303.

Kaseberg, Alice, Nancy Kreinberg and Diane Downie. *Use EQUALS to Promote the Participation of Women in Mathematics*. 1980. Lawrence Hall of Science, University of Californa, Berkeley, CA 94720.

Marolda, Maria. *Attribute Games and Activities*. 1976. Creative Publications, P.O. Box 10328, Palo Alto, CA 94303.

Roper, Ann and Linda Harvey. *The Pattern Factory, Elementary Problem Solving Through Patterning*. 1980. Creative Publications, P.O. Box 10328, Palo Alto, CA 94303.

Seymour, Dale, and Margaret Shedd. *Finite Differences*. 1973. Creative Publications, P.O. Box 10328, Palo Alto, CA 94303.

Summers, George. *Test Your Logic*. 1972. Dover Publications, Inc., 180 Varick St., New York, NY 10014.

Spatial Visualization, Constructions and Coordinates:

Billings, Karen, Carol Campbell, and Alice Schwandt. *Art 'n' Math*. 1975. Action Math Associates, Inc., 1358 Dalton Dr., Eugene, OR 97404.

Brooke, Maxey, See above.

Burns, Marilyn. See above.

Elementary Science Study Teacher Guides. See above. Also, *Mirror Cards* and *Geo Blocks*.

Gardiner, Martin. See above.

McKim, Robert. *Experiences in Visual Thinking*. 1972. Wadsworth, Inc., 10 Davis Dr., Belmont, CA 94002.

Ranucci, Ernest. *Seeing Shapes*. 1973. Creative Publications, P.O. Box 10328, Palo Alto, CA 94303.

Renner, Al G. *How to Build a Better Mousetrap Car–and Other Experimental Science Fun*. 1977. Dodd, Mead & Co., 79 Madison Ave., New York, NY 10016.

Sullivan, Dennis. *Lots of Dots*. 1977. Lawrence Hall of Science, University of California, Berkeley, CA 94720.

Strategy Games:

Burns, Marilyn. See above.

Gardiner, Martin. See above.

Lettau, John, and Bill McConnell, *Primary Dimensions. First Dimension. Second Dimensions*. 1974. L & M Educational Enterprises, Box 88, Santa Maria, CA 93454.

McConville, Robert. *History of Board Games*. 1974. Creative Publications, P.O. Box 10328, Palo Alto, CA 94303.

Also of Interest:

Elementary Science Study Teacher Guide. *Peas and Particles*. See above. (Estimation activities.)

McFadden, Scott. *Success With Solving Algebra Word Problems*. 1978. Action Math Associates, Inc., 1358 Dalton Dr., Eugene, OR 97404. (Guessing strategy applied to algebra word problems).

Perl, Teri. *Math Equals, Biographies of Women Mathematicians and Related Activities*. 1978. Addison-Wesley Publishing Co., Sand Hill Rd., Menlo Park, CA 94025.

Shulte, Albert, and Stuart Choate. *What Are My Chances*? Book B. 1977. Creative Publication, P.O. Box 10328, Palo Alto, CA 94303 (Probability experiments).